Unforgettable 'Memoir'

God remember me.

THE SHERMAN TURNER STORY PART I

SHERMAN L. TURNER

© 2010 Sherman L. Turner. All rights reserved.

No part of this book may be reproduced, stored in a retrieval system, or transmitted by any means without the written permission of the author.

Printed in the United States of America

This book is printed on acid-free paper.

Because of the dynamic nature of the Internet, any Web addresses or links contained in this book may have changed since publication and may no longer be valid. The views expressed in this work are solely those of the author and do not necessarily reflect the views of the publisher, and the publisher hereby disclaims any responsibility for them.

I started praying and begging God to "remember me" because I was so scared and afraid to die.

Table of Contents

DEDICATION IX

INTRODUCTION XI

CHAPTER 1 1
   SEEKING A JOB, 1962–1964 1

CHAPTER 2 10
   PLUMBERS APPRENTICESHIP, 1962–1964 10

CHAPTER 3 21
   APPRENTICESHIPS 21
   1964–1970 21

CHAPTER 4 27
   MASTER PLUMBER TEST AND LICENSE, 1970 27

CHAPTER 5 31
   STARTING A SMALL BUSINESS, 1970–1980 31

CHAPTER 6 41
   TEAMING FOR BONDING, 1980–1985 41

CHAPTER 7 54
   SBA 8(A) CONTRACTING, 1985–1990 54

CHAPTER 8 59
   SBA 8(A) CONTRACTING, 1990–1995 59

CHAPTER 9 68
   FILING OF THE REPORT, 1994–1995 68

CHAPTER 10 73
   THE STROKE, 1995 73

CHAPTER 11 78
   MY FINAL THOUGHTS, 2010 78

APPENDIX A **80**
SPECIAL AWARDS RECEIVED—VARIOUS SBA CONTRACTOR AWARDS..80

APPENDIX B **81**
CONTRACTING REQUIREMENTS THAT SHOULD BE IN PLACE...81

APPENDIX C... **83**
SPECIAL REQUEST—RE-CERTIFICATIONS 83

PART TWO **84**
THE SHERMAN TURNER STORY PART II.................... *84*

CHAPTER 1 .... **85**
THE STROKE, 1995 85

CHAPTER 2 .... **91**
IN THE HOSPITAL: PARALYZED, 1995–1996 91

CHAPTER 3 .... **98**
LEFT THE HOSPITAL, 1996..98

CHAPTER 4 ....**102**
WIFE LEFT ME, 1996... 102

CHAPTER 5 ....**107**
YMCA'S REHABILITATION, 1996–2004 107

CHAPTER 6 ....**112**
MY GUARDIAN ANGEL, 1996–2010................................ 112

CHAPTER 7 ....**115**
UP-SKILLS ACADEMY, 1997–2002 115

CHAPTER 8 ....**119**
BUFFALO STATE COLLEGE, 2002–2004 119

CHAPTER 9 ....**126**

TRIPS TO KENYA, 2004–2006 126
CHAPTER 10 ....**134**
   RACHEL, 2004 AND BEYOND 134
CHAPTER 11 ....**139**
   TRYING TO GET A JOB, 2006–2008 139
CHAPTER 12 ....**145**
   VISITS TO KENYA, 2007–2010 145
CHAPTER 13 ....**152**
   COMPLETE MEMORY BACK, 2010 152
CHAPTER 14 ....**155**
   CONCLUSION, 2010 155

DEDICATION

This book is dedicated to the Plumbers Unions and all workers—the minority male and female black plumbers and all white plumbers, the men and women—who showed strong courage during a difficult time and worked together.

x

INTRODUCTION

My name is Sherman L. Turner; I reside in Buffalo, New York. In 1995, at age fifty-four, I suffered a massive stroke that left me unconscious for several weeks. Since then, I have been undergoing my rehabilitation training.

Because I was 99 percent paralyzed when I came out of the coma, I could not tell my story at that time—I first had to regain my memory, learn to walk again, and finally, regain my speech. When I was in the coma, I remember seeing darkness but no light. I started praying and begging God to remember me, because I was so scared and afraid to die. It gave me the greatest inspiration when I saw a speck of light, shining so brightly. It was like a star floating toward me, but it seemed miles away and within a dark tunnel. Suddenly, the star became as bright as the sun; I could not keep my eyes open because of the brightness. When I woke up, I knew that God had answered me and had given me another chance to live and do the right thing in life.

After more than fifteen years of rehabilitation therapy, I now am able to tell my story, and I shall speak only the truth about the trials of being an honest contractor, an ethnic minority, working for the US government, and a member of the Small Business Administration's 8(a) contracting program. The Small Business Administration (SBA) asked me if I could hire minority workers on government contract jobs. I told them very proudly, "Yes, I can."

Then I learned the hard fact that my contracting partners, who were white and therefore in the majority—I referred to them as the "Wall Street-money contractors"—had other plans. They lied to me and to the SBA about being supportive toward minority hiring and training programs. I had to fight off these contractors, who did not cooperate, regardless of their prior written and verbal agreement with the SBA 8(a) program and me.

The SBA 8(a) contracting program "provides eligible firms with greater access to the resources they need to grow and develop their businesses." I thought that being part of this program would bring my dream of success. Instead, it was the beginning of a huge fight over hiring minorities on two jobs that for which I contracted with the government at the Veterans Affairs Medical Centers (VAMC) and the West Valley Nuclear Services. Yes, my contracting partners—the white Wall Street-money contractors—did not like my hiring and training minorities. They wanted me to hire non-minorities, and they did not want me to provide training or jobs for minorities.

An ugly misunderstanding may have existed between me and my white contracting partners with regard to my agreement and contractual obligations with the government. But the misunderstanding was not mine; I always was very clear that my being the government minority contractor meant that I would make every effort to hire and train minorities. I think the Wall Street-money contractors, who were worth millions or billions of dollars, resented my hiring minorities, as well as resenting the government's wanting the hiring and training of minorities.

I am so grateful to the following wonderful people: Sabrinna Turner, La Vonda Bailey, Bertha Turner, Paulette Turner, Richard Bailey, Willie Roberson, Shaun Roberson, government SBA officials, and all those tremendous minority organization owners who help make having an equal opportunity a reality. Special thanks to the SBA government officials in Buffalo, New York; the University at Buffalo; the YMCAs in Buffalo; Buffalo State College; Erie Community College; and my Kenyan rehabilitation school, including Bwana Chango, Mwalimu Iha, Bibi Onzere, Bibi Onyango, and my young men in Kenya: Timo, Nelson, Michael, and Jimmy. Also, special thanks to the friends of President Obama in Kenya, who gave me special help and instructions in the gym with my rehabilitation workouts.

CHAPTER 1
SEEKING A JOB, 1962–1964

I grew up in a family that was very poor; we lived in the ghetto. Schools in the ghetto at that time were substandard—the teachers didn't want to be there, and they didn't really care about teaching us anything. For example, for the end-of-the-year testing in the first grade, my classmates and I were told to write the letters of the alphabet. I raised my hand and told the teacher, "I don't know how to do that. You never taught me that!" My teacher answered, "Look up on the board around the room and just copy the letters, and shut up." Yes, in almost all ghetto schools, children would just copy from the board during the final tests.

During my high school years (1955–1959), it was drilled into me that I would not be able to get certain types of jobs after graduation, such as being a lawyer or doctor, because it was "not a black man's type of job." The guidance counselor would always recommend that black students seek a factory job. My mother, however, encouraged me to learn a trade. She said, "When the factories won't hire you because you're a black man, then you can start your own business and have a job, regardless of what the white man says." Plus, she said that because I had no possible way to get a college education, I must try to learn a good trade.

I wanted to do something for my mother. She needed help taking care of my brothers and sisters, so she eventually agreed that I could join the army. This seemed a better option than my trying to find a job—no one would hire me because I was black—and it had the additional advantage of providing me with health insurance, which was often denied to black men in their jobs. My mother knew that serving in the army meant that I would get lifetime insurance coverage through the

Veterans Health Administration.

My tour of duty in the US Army was from 1960 to 1962. The army sent me to Louisville, Kentucky, arguably the most "Jim Crow" city in the United States of America. On my very first day in the army, I was placed in the holding detention center after I drank from the "wrong" drinking fountain and used the "wrong" restroom at the train station. In those days, drinking fountains and restrooms were segregated, clearly marked as "Whites Only" or "No Coloreds." A military officer asked me if I'd read the sign on the restroom—"No Coloreds"—and said, "That means you." I asked, "Why me? My name is Sherman, and that sign did not have my name on it." Then another officer said, "Boy, do you want to get killed? You must not know where you are now." These army officers then looked at my train ticket and saw that I was from New York. One looked at the other and said, "He's one of those Yankees. We should call his mother; he doesn't look like eighteen years old anyway." After calling my mother, the officers came back shaking their heads and saying to each other, "He really does not know the rules of where he is now."

I remained locked up for several hours, like a prisoner, all because I'd taken a drink from the wrong drinking fountain and pissed in the wrong restroom. Later, the army officers introduced me to Sergeant Crawford. They said that he was responsible for my actions and that I should listen to him so I wouldn't get killed while in the army. Sergeant Crawford looked at me and said, "Boy, it's me and you now, and I own you. I've already talked to your mother, and now I am your mama!"

The next morning, Sergeant Crawford came to the barracks where I was staying and began teaching me the rules of the army and the way things were in the Jim Crow South. In the mornings, my lessons were on boxing and wrestling—this was because everybody wanted to fight me because I was so small. At eighteen years old, I weighed not quite eighty-five pounds. In the afternoons, the hard training started; it detailed how blacks should behave if they wanted to survive in the South.

After a month of this training, Sergeant Crawford told me, "We have to start double-timing your training before basic training ends." I then realized I had been missing my army-required basic training classes and courses while Sergeant Crawford was instructing me in his own way. One day after that, he said, "Boy, don't worry about your basic training courses; you're not ready yet. But in next four weeks, you will be ready to survive in the army."

I never had army basic training, but I did have basic training on how to survive in the Jim Crow South, which for me, coming from a northern state, was hard. Soon after, I was granted military leave and went to Buffalo. At the train station, I saw the same white military officers who had arrested me on my first day—and saved my life. I smiled and gave them a big salute. They smiled and said, "Sergeant Crawford did an excellent job."

When I got home, my mother told me that when Sergeant Crawford called her when I first started my training, she told him that she was Papa Richardson's daughter from Arkansas—she herself had been born in Buffalo because her family had escaped slavery and fled to the North before they could hang her father. She remembered the Crawfords from Papa Richardson's stories, and she asked the sergeant to protect me because I did not understand about the ways of the Jim Crow South. Sergeant Crawford told her not to worry. "I will train him," he assured my mother, "and prepare him so that no one will hang him from a lynching tree." I was fortunate to have received the special training from Sergeant Crawford. During my time in the army, I learned that the foundation of racism was deeply imbedded in the army. The whites carried their hatred of blacks everywhere, even to countries overseas. I knew this because when I was in other countries, the people would ask me, "Why do you want to live in America, where you cannot get a job because you are a black man?"

I was assigned as quartermaster in the army, which meant it was my job to load and unload trucks. When I tried going to school while in the army to get my GED certificate, the army's response was to give me four weeks of KP duty (work in the kitchen). I soon learned that I faced the threat of being placed in the stockade if I tried going to school, so I escaped regular army duty by accepting the temporary duty assignment, or TDY, for basketball, boxing, and running track for all but three months of my two years of duty.

After completing my tour of duty in the US Army, I was ready to seek a job in the real world. I was hopeful that times had changed for the black man, but when I started a job search, I was told, "This type of job is not for a black man. Try the steel mills." What that meant was that *any* type of job was not for a black man. When I tried to get a job at the steel mills, however, I was told that I had to weigh at least 160 pounds, and I weighed only 120 pounds. I know now that I was being given the run-around because they didn't want to hire me—that was their way of saying, "We don't hire black men."

One day I went looking for a job at the employment office and saw a poster that indicated the Plumbers Union was seeking apprenticeship trainees. I went to the Plumbers Union but was given a similar excuse—their tradesmen jobs were usually for white men only, but I could always come back and apply at a later time. As I was exiting the building, a man I'll call Mr. Jenkins approached me, asking why I was applying for a tradesman job. I told him that I'd just returned from defending our country in the Berlin Crisis in Germany, and my mother had told me that being a tradesman was the best type of job for a black man who wanted a good future. Mr. Jenkins told me to come back in and reapply for the apprenticeship training.

Mr. Jenkins told the other members of the union that he was accepting me for the five-year training program and giving me a five-year probation, subject to my marks, which would have to be passing to remain in the training program. His face then turned a fiery red as he angrily told the union members that he remembered how the Plumbers Union had discriminated against him and his brother because they were Italians; therefore, he refused to discriminate against the blacks. It was clear that Mr. Jenkins was boiling mad as he told them, "If you don't like it, fire me!"

I knew what a blessing it was to have the opportunity to learn a trade. I had to study harder than ever because I knew of my shortfalls; I had not received a quality education in the ghetto. I said to Mr. Jenkins, "Would you please give me extra work assignments to do at home so that I can learn and keep pace with the others? I do not have a background or any family in the plumbing trade." Mr. Jenkins agreed.

He also asked the other teachers to give me extra-hard assignments to help me learn and said that they should explain things more in detail so that I could grasp the concepts better.

CHAPTER 2
PLUMBERS APPRENTICESHIP, 1962–1964

The Plumbers Apprenticeship program required classroom work two nights a week, for five years, while working in the field eight hours per day. Almost all of the white guys had jobs with their fathers' or other relatives' companies. In my class, I was the only one who did not have a job. Most of my fellow students were eighteen to twenty-one years old, but because I had already served my military duty, I was older, at twenty-four years old. The real difference between us, though, wasn't our age; it was that I was a black man.

Mr. Jenkins did warn me that because I was the first black man in the program, just as he had been the first Italian man, there would be resentment from my classmates. He said that he knew nothing had changed; there was still discrimination, because they also resented his being the plumbing school director.

One student in my class had a plumbing engineering degree. The instructor said that this student should be teaching him! I wondered why this student was in class when he already was an engineer, but I found out that he could not get his master plumber's license without serving an apprenticeship, even though he was running his father's plumbing company and was the plumbing estimator for the company.

The course work was hard for me because everything was new, including the plumbing terminology. My schooling in the ghetto did not prepare me to compete against these other students, and realizing this fact made me angry—I knew it was not my fault, but I felt like I was the dumbest student in class.

During the five long years of schooling, some guys tried to become my friends, regardless of the fact that I was a black man, but most were racists—this was the early '60s, and the whole Plumbers Union was racist. It surprised me, though, that all the instructors were just as racist as the students. They seemed to enjoy giving me those hard homework assignments and would laugh about it. But I knew that I needed the homework assignments in order to do well. I studied hard, and soon, my marks were 100 percent on every test—and then the white students noticed me. Even the student with the college degree who was the plumbing engineer noticed my school marks.

On the day we took the final test, all the other students said that I would not get a better grade on the test than the guy with the college degree. Throughout the previous month, however, I had studied really hard for at least eight hours each day. Therefore, I confidently told the whole classroom that I would score better than the plumbing engineer.

After the instructor scored the test, my mark was again 100 percent. The other students never showed me their school marks. They just stopped speaking to me, and I knew that it was because they felt safer reverting to their old discriminatory behaviors.

I wondered if maybe that was why nobody would hire me—because my passing marks in school were the highest of all the plumbing students. I also wondered if that meant that nobody would *ever* hire me.

At night when I came home from school, my mother would always try to pump me up, emotionally, because she knew I wanted to quit school. Actually, I just felt it was a hopeless situation. Racial discrimination made life very hard, and I often thought it would be easier to just quit. But Mama used to say, "Do your best in everything that you do and someday you will get a job."

The time in plumbing school training passed by quickly, and soon those five years were over. During that time, no plumbing company had hired me to work. All the white plumbing trainees had been working for about five years, so now, with five years of schooling, they could become full journeymen plumbers. But I was still an apprentice without work experience—and I apparently could not get work experience because as a black man, no one would hire me.

Mr. Jenkins told me that my grades had been the highest in my apprenticeship class, but even so, he was sad to say that the plumbing board had chosen a white guy—second place in our class and whose father was a big-time master plumber and had big-time money—to represent the union in the apprentice contest. He said this was the first time ever that the student who was second in the class was chosen to represent the union, which was an honor. I was not even acknowledged, even though my grades had been the highest.

After waiting almost five years, a plumbing company finally hired me, and I started my on-the-job training (OJT)—I needed to get at least five years of field-training experience to become a journeyman plumber. You see, the Plumbers Union sends its members to work for its membership plumbing companies, and then that plumbing company pays the worker. Members of the union may work for many different companies but only one union.

That first plumbing company that

eventually hired me had tried to keep me out of the field by telling me their field plumbers were just too rough for me. I'd been in the army and was fully trained in using weapons! Why would they say that the plumbers were "too rough" for me? I believe now that it was a scare tactic to get me to quit. The next plumbing company I worked for was different; I worked in the field with other plumbers, but they never talked to me and never ate lunch with me. For years, nothing changed; the percentage of white plumbers on the job was 100 percent—until I got there. The white plumbers only spoke to me when they wanted to use me as their coffee boy and flunky.

At a company Christmas party one year, the white plumbers and white apprentices drank too much and then ordered me to shine their shoes or to perform other flunky-like tasks to humiliate me. I decided to never again attend another party where I'd be subjected to such treatment. One thing did surprise me—I heard one woman at the Christmas party tell her husband that he was embarrassing her with his obvious racism and that he should just take her home. Still, everyone else either joined in or ignored the discriminatory and embarrassing remarks.

The next time I was invited to a company Christmas party, I tried to excuse myself by saying that I was busy that day, but I was informed if I didn't go, everyone would think that I didn't like white people. Therefore, I felt I had no other choice. Early on, the party actually was fantastic, and I was surprised that I really enjoyed myself. And then the drinking started; I did not know that they had spiked the punch. My job foreman said to me, "Sambo, come shine my shoes and dance for me." I felt like punching him in the mouth, but I knew that was an immature reaction that would have cost me my job. Instead, I quickly made my exit. The next day, my foreman tried to excuse himself by telling me that he'd been drunk, but I said, "I am transferring to another job. This is my last day with you."

Later that day, an apprentice friend of mine who happened to be working nearby came to my job to eat lunch with me. I was surprised when my foreman approached him and demanded to know what my friend was doing there. My friend just glared at the foreman and answered, "Not minding your own business on our lunch time is a good way to get your behind kicked."

I was shocked that my friend had been so bold. "Why did you say that to him?" I asked.

My friend answered, "That plumber has a reputation for picking on smaller black guys in the union. I had to show him that he doesn't scare me." His plan apparently worked—when I saw my foreman the next day, he didn't make any racist remarks to me, as he usually did. Even the other workers on the job noticed that my foreman had toned himself down and was no longer so openly rude. I would not have complained to my friend or anyone else about the way my foreman had treated me, but I was glad that my friend had spoken up to him. That experience taught me that we are our brother's keepers, just like the Bible says.

From 1960 through 1966, older black men tried to learn the construction trades through Project Justice—a ninety-day federal program intended to help minority workers—but the white plumbers resented the black men who tried to work on the job. The white plumbers said they would not help the black men learn anything. I always felt that the white plumbers simply were jealous that these men could go through training in ninety days. Granted, this training period was much shorter than the white plumbers' training, but these older black men had been discriminated against for many years; they never had an opportunity to work because no one would hire them. Project Justice gave them a better chance to earn a living, yet the white plumbers still resented them.

I once saw a black plumber who was part of Project Justice start to light his torch incorrectly. I told him to stop, saying, "You will blow yourself up! Please don't do anything before I check on you. I am here to help you learn to be safe." All the white plumbers got mad at me for helping this man, as if I should ignore the fact that the man hadn't had proper training.

My cousins and others who were part of the program told me that their class instructors had limited knowledge of the construction trades and limited working experience. I also learned that their school books were over twenty-five years old and often missing pages. Many times I wished I could have trained those men myself. Instead, I was determined to watch out for the many plumbers who came through the Project Justice program and to treat them with respect when I was a contractor. I thought of the program as Project *Injustice*, partly because of the injustices shown to minorities over hundreds of years, and partly because these minority workers needed training that they never received.

The attitude of the white guys on the jobs was often unbearable. Just being in their company and witnessing all their hatefulness made me so angry that at times, I felt like quitting.

"That's exactly what they want you to do," my mother would say. "Don't be a quitter."

I would answer, "But Ma, when I see how our cousins get treated like dirt and without respect on construction jobs, it hurts." I knew she understood this, but I never said anything more; I knew better than to give Ma any back talk.

Malcolm X once commented that when discrimination hits you in the face, it hurts more. After leaving the army and working in a field that was almost entirely white, I understood exactly what he meant. I also understood Muhammad Ali's comment that black men in America do more fighting by trying to get a job than white men did in fighting in the world wars.

When Malcolm X was killed in 1965, our house and our hearts were silent, because Malcolm had been our truth and soul in those days. Malcolm was the driving force for the black man working in construction jobs and for those wanting to own businesses, too. Malcolm X stood up for his people long before civil rights became the law. People questioned why he did the things he did. It was simply because it was the right thing to do.

CHAPTER 3
APPRENTICESHIPS
1964–1970

I continued working as an apprentice for the Plumbers Union until no one would hire me anymore. This was because fifth-year apprentices made almost the same hourly pay as a journeyman plumber, so many of the white plumbing companies requested that the union send minority first-year apprentices only. This resulted in my being laid off for over a year, without hope of getting a future job. In the Plumbers Union, a black man was always regarded as inferior to a white man; the black man was never given an equal opportunity.

My mother suggested that my only options were to go into my own business or get on welfare. We had a big laugh when my mother mentioned welfare, because I'd joined the US Army to help my mother get *off* welfare—she was still trying to get off welfare and get a job.

I chose to go into my own business. That meant I would have to take the Master Plumber test. At that time, the city of Buffalo had never had a black master plumber and did not allow black men to take the test until 1970. I approached Mr. Jenkins and asked if he could help prepare me to take the city of Buffalo's Master Plumber test, because no one would hire me as a black plumber. Mr. Jenkins said that he could not teach me anymore. For five years, he and the other teachers had given me extra-hard homework assignments to prepare me for the Master Plumber test. He said they all had known that this day would come.

Mr. Jenkins told me that the city's next Master Plumber test was the following month and that I should mail in my application—he suggested that I mail it so no one would know that I was a black man. My teachers and Mr. Jenkins also advised me to take the test in ink, not pencil. Historically, Buffalo was a strongly racist city. Mr. Jenkins said that if I answered in pencil, someone could erase and change my answers on my test papers. The very next day I filled out my application and mailed the test fees into the city of Buffalo.

A man from the Plumbers Union called me and said that he'd heard I was planning to take the Master Plumber test. He said, "If you do take the test, we will have to expel you from the union, because your taking the test is against our rules." I responded, "If you can put me to work, then I won't take the test." He snapped at me, "Don't talk smart, you nigger, because you may lose more than your job!"

I tried not to let his threats affect me. I had to prepare myself, physically and mentally, for the Master Plumber test. I made a schedule for studying four hours a day, as well as exercising by running several miles per day. Each day as I set about my tasks, I would think of my mother, brothers, and sisters, and I felt deep empathy for those black men who were discriminated against and denied the opportunity to earn a livelihood. It was around this time that the deaths of

Malcolm X and Dr. Martin Luther King Jr. caused more racial tension. I received threats against my life for taking the Master Plumber test. That was crazy—how was I harming anyone?

I needed time to get pumped up, so I could represent all those other black men who'd come before me and were denied an opportunity. My plan was to "smoke" this test and represent all my people proudly, but on the day of the test, I was very scared. I thought about my life as a poor black man with no future, someone whose life was threatened just for trying to take a test that would allow him to work and help his mother and family.

I was very nervous, thinking about that threatening phone call from the union. The test started at 8:00 a.m. and continued until lunch, which was from noon to 12:30, and the test ended at 4:30 p.m. As I walked into the test room, all ninety-nine applicants looked strangely at me—I was the only black man taking the test. Their collective look seemed to ask, "What are you doing here?" Then I gazed across the room and saw that Mr. Jenkins and my schoolteachers were test monitors ... but then I noticed that the Plumbers Union official was behind me, taking the same test. I remembered how Malcolm and Martin had died because of hatred, so I changed my seat so that I could face the Plumbers Union official. *I am a man and not afraid of anything*, I told myself. The union official never looked at me.

Because I knew that I had to do the test in ink—so that no one could change my answers—I planned to read the questions and answer slowly. I did feel very relaxed now that my enemy was not sitting behind me. The test began ... and by about 10:00 a.m., I was finished with the eight-part test and the drawing part. Just as my teachers had told me, this test was like my extra-hard school assignments. I looked across to see if the Plumbers Union official had finished the test, but he was still on part one, with seven other parts and the drawing yet to complete.

I raised my hand for the test monitors to take my completed test. Mr. Jenkins asked what I thought about the test and if I thought I'd passed. I smiled and said, "The test was pretty tough, but I think I got 100 percent." The test monitors then checked to see if I'd answered all eight parts and completed the drawing. They told me that I could go because I was finished.

The union official looked up and shouted, "How can he be finished when I am still on part one? He must be some kind of genius—or he cheated." As the union official shouted his remarks, I could hear all the pencils in the room drop, and then there was silence, and all eyes were on me. Right then I knew that my plan had worked—all ninety-nine white applicants knew then that I represented all black men and that I had "smoked" the test. I laughed and quickly left the room.

I found it amazing that God had allowed me, a black man from the ghetto with an eighth-grade education and a high school GED, to pass that plumbing examination. Yet something strange had happened during the exam: when I read a question, all of a sudden it was as if there was a picture in my mind. I could easily answer all the questions because it was as if I was copying the answers from my mind.

CHAPTER 4
MASTER PLUMBER TEST AND LICENSE, 1970

After several weeks of waiting for the official test results, I was informed that I had not taken the Master Plumber test because no test records were available—or at least no records could be found. I told Mr. Jenkins about this, and he said that I would have to see the mayor of the city of Buffalo, because my papers and license were missing from the City Hall safe. He suspected that someone in the Plumbers Union—and he thought he knew who he was—had taken them. Clearly, because it hadn't been possible to erase my answers, my papers and license had been stolen.

I made an appointment to see the mayor, and when I went into his office, he greeted me very nicely. He said that he'd called the Plumbers Union and told the person—the one whom Mr. Jenkins suspected—to find the city's missing property and to return those items and my license by 9:00 a.m. tomorrow. I was to report back there the next morning at 9:00 to meet Buffalo's Common Council... because I was the city's first black master plumber.

The Plumbers Union did expel me from the union for taking the test—they said I had disobeyed them by taking the test; also, the Plumbers Union official had failed the same test on which I had earned a perfect score—that did not sit well with any of them. I also lost all my benefits—this after working almost five years with the Plumbers Union.

I became a contractor because at that time, the unions and businesses would not hire black men. Unfortunately, most white men had a false notion that black men were inferior. My mother had been right—I would never get hired in the white man's world. Becoming a contractor was the only way I could find work.

Every time I went to a new town or village on a contracting job, I had to take the Master Plumber test again; each location in the suburbs had its own test for licensing before someone could work in the community. In one town, when I asked the chief plumbing inspector for the application to take the town's Master Plumber test, he said, "You must be that guy, Sherman Turner." Then he told me no, I couldn't take the test, adding, "Not as long as I am living. You can come back when I am dead and in my grave! Now get the hell out of my town!"

I came to realize that this type of attitude was to be expected, because discrimination and racism existed strongly in the suburbs. Even so, I had many white customers in the suburbs, and they all paid me very well. As a matter of fact, in the town where the inspector had told me to come back after he died, several of my customers worked in that inspector's office. It became an inside joke that some of his own office clerks were my plumbing customers.

Within three years, that plumbing inspector died, and I went back to the town to ask for an application to take the Master Plumber test. The clerks at the counter said. "We've been waiting for you to come back. We have your application right here. Please take the papers and turn them in to the town clerk with the fees." Then all the clerks, secretaries, and people in the office came up and shook my hand, saying, "We are glad to meet you, Sherman."

When I applied at the office in another town in the suburbs, the clerk asked, "Are you Sherman Turner?" When I said yes, the clerk said that I wouldn't need to take the test for them; they recognized the previous town's license and testing, so they would give me an honorary certificate license at no charge. I was pleased that they knew my name, and I thanked them very much.

There were mixed reactions in the other towns and villages where I applied for my Master Plumber test, but one thing remained the same: in every location there had never been a minority in those types of jobs. Every year when I would renew all my master licenses, I always would say, "This place needs some color because it's too bright!" One town actually hired a minority, so when I walked in to renew my license, I said, "Your office has nice color now!" Everybody laughed and said, "You know, Sherman, you are right."

CHAPTER 5
STARTING A SMALL BUSINESS, 1970–1980

Life as a black man meant adjusting my survival tactics, and starting my own business was a means to survive. Starting a business meant I had to get a truck and tools, which I didn't have. Plus, I had no job and no income, remember, because the Plumbers Union had kicked me out the union and had taken away my benefits. I looked in my pocket and found exactly thirty-nine cents; my wallet was empty. This made me feel like crying, but I kept my composure because my thoughts were traveling a mile a minute.

When I was younger, I'd had a paper route, so I now decided to contact the customers on my old paper route and let them know that I was a plumber; I asked them to please support me in my efforts. They all were so happy for me, and the community did support my new plumbing company, which I called S&S Plumbing. I started several jobs the very next day. I also did small jobs for the elderly in the community at no charge, because they did not have the money but needed the work. Soon, the elderly, as well as their sons and daughters, started calling me to do their work, and my business just took off. I felt proud when the elderly offered me fish and chicken dinners as payment—the food was delicious! And these folks really spoiled me with cherry and apple pies, too.

The "S&S" in my company's name stood for "Sherman and Sherman"—my uncle's name was Sherman, too. Uncle Sherman was the community undertaker, and he ran a successful business for over fifty years. As my uncle had done in his business, I bought church fans to advertise S&S Plumbing. I wanted to emulate my uncle's success.

One day while I was visiting Uncle Sherman and his wife, my aunt Bessie, I noticed that their bathrooms were out of order and asked what the problem was. Aunt Bessie sighed and told me that the man who was fixing the bathrooms said he was waiting for parts. I went to my truck and got my tools and extra parts, and I fixed the float valves and installed new a washer in the lavatory faucets. My aunt Bessie was amazed and very happy.

My uncle Sherman was someone I could be proud to emulate because he was a wealthy businessman, councilman, undertaker, and head deacon in church, and he always gave to the poor. During the Depression he would set up soup lines in the poor community on Buffalo's east side. Usually, Uncle Sherman didn't talk much to us "kids," but one day, after I started my business, he called me. "Boy, I hear good things about you!" he said. Then he asked, "What's the name of your plumbing company?" I told him, "S&S Plumbing—that stands for Sherman and Sherman, because one day we will be partners, when I am rich." My uncle Sherman said, "You're my boy, and I'm watching you."

When there was no plumbing work of any kind, I had to look for a second job to supplement my income, just to be able to eat. I applied for a plumbing inspector job, which was funded with the government's CETA (Comprehensive Employment and Training Act) funds. It paid five dollars per hour; the regular plumbing inspectors got twenty-five dollars per hour, with CETA funds. I was told, "You can have the job if you want, but we cannot pay you more than what the government gives us." I accepted the very low wages because it was better than nothing.

After I'd worked for about a week, the big boss—the chairman—came to see me. As it turned out, it was the man who had been one of my favorite plumbing teachers. He immediately assigned me to review and approve blueprints and check for correct installations. I knew that my training in plumbing engineering was just beginning. Now, I was getting free training and learning engineering. God was great to help a ghetto guy like me get an education from the best teacher. This engineer told me to use my photographic memory to visualize blueprints, and he taught me things I couldn't have learned in school. After his training, I knew that I was the "real deal"— a master plumber.

My special abilities as a plumbing inspector were an asset to the town, which was a mostly white and affluent area. The older plumbing inspectors never learned about blueprints and hadn't dealt with federal, state, or local town codes. Plumbing and piping and all codes were my specialties, as was using my photographic memory. My assignment was to check out unauthorized activities and illegal acts in plumbing and construction work.

One day, I wanted to check a house that looked very suspicious to me. I'd seen an unmarked truck hidden at the back of the house, with unmarked excavation around the building. I was shocked when I saw that the owner of the house was a black man, someone I had seen in school. He was living in this fabulous, expensive house in a very wealthy white-people's town. I could not say "Hi, bro" to him, but from what I observed, this brother was getting ripped off by a white plumber. I had to remain cool so the brother would not get upset.

I approached his plumber and asked him to give me a tour of what he was doing at the house. The plumber showed me his materials, which were illegal (he was using type "L" soft copper in the ground; it should have been type "K" copper), and then showed me his job layout. The layout also was illegal because code dictated that all fixtures had to have a vent piping to the ventilation plumbing system for proper flow; his did not. And he showed me his tools, which were the wrong tools for that type of job. I asked him for his permit, and he told me he didn't have one yet. He said, "Your boss, Mr. Lachut, told me I could get the permit later. I called it in to the office." The chief plumbing inspector always instructed me not to listen to the lies that plumbers would tell when they got caught working without a plumbing permit. To do my job to the letter of the law, it was my responsibility to call the police—and that is exactly what I did.

When I got back in the office, I told my boss about what I had seen. I said, "Mr. Lachut, why do the white people in this town get a quality plumbing job, but this honest, hard-working black man gets ripped off? The home owner had paid the plumber for his work, but that man was no more a plumber than my baby finger. I did my job; now you make it right, because that black man has no more money to pay."

Mr. Lachut nodded in agreement. "Don't worry, Sherman. I will make it right."

The next day, all the inspectors called me "Lachut's hatchet"—it was my new nickname. Later, as my boss spoke to other white plumbers who claimed they were ready for inspection, he asked them, "Are you sure you're ready? Because we may send the hatchet man. Are you sure your plumbing installation is correct? You know what happened to the last white master plumber that the hatchet man caught!"

It mattered little to me that I had called the police to arrest a white master plumber. At the time, all I saw was someone who was being ripped off by a crooked plumber. I then remembered the time my mother was depending on a plumber to give her an honest job, and the plumber ripped her off, too. I didn't care if anyone called me "Lachut's hatchet"; I would not allow that brother to get ripped off. In fact, I welcomed the nickname because to me, it indicated honesty.

When the plumbing repair work slowed or stopped altogether, I needed to find other sources of work. The Small Business Administration (SBA) requested plumbers who were willing to work in the national flood hazard areas in Elmira, New York. One day in 1970, I went to the SBA to see how much they wanted plumbers—that is, if black plumbers were okay with them. They told me that they would take anyone, so I rounded up four men who needed a job, and we went to the flood area in Elmira. When we reported in, the supervisor sent us to the worst area in the black section of town. The area was devastated—completely torn up—and black folks had no place to live and had no shelter.

I thanked God that I had brought a new hydraulic pipe pushing machine with me, because what they needed were new services of water and gas installed in the temporary house trailers they were using for housing. A hydraulic pipe pushing machine saves time because it doesn't excavate the complete length of an underground piping installation, which would use many manpower hours of digging and backfilling. Instead, a hole is dug in one spot and then, fifty feet away, another hole is dug, and the pipe is pushed from one hole to the other hole. This saves time because no excavation is required.

Because I had the correct machine, I could set up each man and teach him how to use the machine, so we would have maybe one or two hours of digging. Without the correct pushing machine, we would have had to do complete trenching and backfilling, which takes two or three days per one complete unit.

After I taught my four men how to use the pushing machine, we completed four to six units per day. Plus, I was an expert in surveying layout and was schooled-trained with the transit level. A transit level is a surveying instrument used to measure the height and depth of underground piping coverage. The white plumbers were completing only one or two units per day and had no transit level. The Small Business Administration noticed that my men and I worked faster; the SBA paid on a per-unit basis, and per week, my billing was for thirty or forty total units. When we finished completing the black area, the SBA asked me to stay and work in the white area, mainly because the SBA engineer didn't trust anyone with putting in ground work without a transit level.

My crew and I stayed an extra two months, helping out the SBA. When I was getting ready to leave, the SBA asked me if I knew about the SBA 8(a) Business Development Program and said they would recommend me for it. I agreed because I needed a job, even though it was the first time I'd heard of this program.

I soon learned that the SBA 8(a) Business Development Program was designed to assist in the development of small businesses that are owned and operated by individuals who are at a social and economic disadvantage.

CHAPTER 6
TEAMING FOR BONDING, 1980–1985

As a contractor, I needed to be bonded. A contractor's bond is a type of financial guarantee that promises that the contractor will finish a job as expected. If he fails to do so, the agency that issued the bond then provides whatever pay-out is necessary for compensation. SBA contracts and all government jobs required bonding, which put minority contractors at a great disadvantage. This was certainly true in my case; I could not get bonding. My only option seemed to be teaming up with a larger company—one that had no trouble getting bonded; one that was run by the white men with plenty of money, or as I referred to them, Wall Street-money contractors. Before I did so, I decided to have a family meeting with my brothers. I knew that they would never steer me in the wrong direction. I made my presentation of the benefits of teaming with the white contractors, not the least of which was the opportunity of getting government contracts.

But my big brother, Donald, said, "You fool. Have you forgotten that white men always lie when money is involved?"

My younger brother, Gary, agreed with Donald. "They just want to use you until they don't need you anymore, and then they'll throw you away," he laughed.

I then told both brothers, "Why do we have to be at war with the whites? They all cannot be liars, right?"

Donald said, "How can you trust them? We know you want to hire and train minorities, but you know that sooner or later, the other contractors will change their minds and call it business."

"But this time the SBA said they will be my witness to verbal and written agreements," I argued. "And the SBA also promised not to allow any type of discrimination on federal government contract jobs. If any agreement is broken, I will take it to court. That's my promise to you brothers."

Donald said, "Okay, sounds good to me." Gary said, "Sounds good to me." And I said, "You both have my promise."

It was in the early 1980s that I teamed up my small firm with a larger company, strictly because I could not get bonded on my own. The SBA needed to approve such a partnership, however, because sometimes a white contracting company would use a black contractor as a "front" to give the white company the appearance of hiring and working with minorities. I brought in the principles from the white contracting company, along with the company president, and we met in the SBA offices.

After several meetings with the SBA and the bonding company, the SBA was convinced, as I was, that I would not be a black front for the white company. These white contractors needed work but could not get SBA government contracts; my firm could get the contracts but first I needed bonding. So it was agreed that the Wall Street-money contractors would bond my firm, and then I would subcontract some of my jobs to them. I would then do the rest of the jobs with my own men. Everybody would get a piece of the pie. It actually was a win/win setup, but if I hadn't agreed to it, I would not have gotten any work. I would have preferred to work without the white contractors' involvement, but at that time, the team arrangement was better than nothing.

The white company with which I teamed was not a family-owned business; it had various shareholders and various vice presidents, too. The agreement was that before the company could use me as their minority contractor, I could borrow fifty thousand dollars from them to use as my payroll (to pay myself), while they taught me estimating and how to manage a profitable construction business. This was acceptable to all parties. The bonding company, my company, and the large white-owned company with which I teamed all signed the agreement.

These agreements and the training were my insurance that the larger company would treat my smaller company with fairness and respect. At that time, my greatest concern was the ability to run a legitimate minority company.

The president of the white-owned company told me that I would have to get my estimating and management training at their Niagara Falls office, because their Buffalo office was not ready for a minority person yet. He smiled and said, "I have to give some people more time to accept change."

I agreed, although I thought the president's comment was strange—just that morning, I'd met some of his people, and they seemed nice and pleased about the fine opportunity for me. Several people had already made plans to start tutoring me in Buffalo. It made me wonder who really was "not ready"—the president or his people.

Additionally, because white contractors wanted to use me as the in-house minority contractor, the Plumbers Union, which had expelled me and had taken away my benefits, suddenly restored all my earned benefits and gladly took me back into the Plumbers Union—it was because I was in partnership with a white contractor. As a minority contractor, everyone looked to me to lead in the hiring of minority workers; the union had not addressed this issue appropriately. In 1980, there were fewer than four black plumbers in the union of one thousand plumbers—that was shameful. It was agreed that if the union could not properly supply me with men, I could ask unemployed minority workers to join the union.

My being in the SBA 8(a) program meant that the education I had been denied in my early years now was available to me. The SBA personnel recognized that early on, so they signed me up for several computer classes. I was very lucky, because the Buffalo SBA had great personnel. I know that I must have seemed to be the most illiterate person to them, but they were surprisingly helpful to me. The special counselor assigned to me allowed me to take mathematics classes and computer classes—I didn't even know about computers at that time.

The SBA personnel all had college degrees, and they said their grade point average in college was 4.0. It was many years later, when I started attending night-school classes in college, that I found out that a 4.0 grade point average was the highest mark. When I think back on the SBA, I see that they took a backwoods plumber into the computer era. All the things I wanted to learn became clearer, because the SBA personnel took the time and also allowed me to attend their in-house classes.

I personally had a difficult time understanding the issues relating to bonding—maybe that was because it belongs more in the insurance field, and many poor ghetto people, like me, never get that far into education. Soon, though, thanks to the SBA, I quickly became an expert student, with knowledge of bonding. Still, the SBA personnel never offered me any job opportunities while I was attending classes. I did not complain, however, because I knew my educational abilities at that time were limited. After the SBA finished with my training, my job opportunities exploded and became unlimited in the construction field.

The next big issue that the Small Business Administration helped me tackle was how to market my company and business. At first, I did not know what these college folks were talking about—what market? I went to the market to get my groceries. Then they enlightened me that marketing is essentially advertising. My counselors enjoyed helping me learn about these new ideas—and they never laughed so much as they did with me. When I was approved into the SBA 8(a) program, I was really nervous. I wondered just how these people who had the good life and a college education would relate to and understand my ghetto-type language. I needn't have worried; soon, the SBA offered me a contract job with the Veterans Affairs Medical Center (VAMC) hospitals. All I had to do was estimate the job and then compare it to the government estimate to see where there might be any differences in costs.

It's a good thing that I had excellent experience with computer estimates, because the government's estimate was the outdated type of handwritten spreadsheet that was hard to decipher. Again, I used my photographic memory in the conference room because I had no blueprints with me. I told the engineers where to look for the differences, after which we negotiated an agreement for an increase on the contract price. The old price was $200,000; the updated price was $350,000.

One advantage of being in the SBA 8(a) program was that I could get free contracting lessons on various subjects, such as bonding, advertisement, marketing, and accounting, among many others. Some topics required a minimum fee, but that still seemed like they were almost free. I knew nothing about contracting until my SBA 8(a) experiences. SBA officials also were present in many of the classes I attended, making it easier for the class to discuss different aspects of the various subjects.

One of my jobs was at one of the old steel plants in Buffalo. That seemed to be the turning point, where the use of my photographic memory helped me excel in my job. The job involved replacing the existing drain lines and water lines to all plumbing fixtures, as well as the equipment to machines. Because this work was performed while the steel plant was in operation, exact planning and coordination was critical.

I was just a small contractor, so I had to work on this job myself. The complexity and magnitude of this small job, however, turned it into a huge project. The steel plant suddenly wanted the whole plant redone because it was advantageous to their taxes. Instead of my three plumbers, as I'd originally planned, I had to hire twelve plumbers. Can you imagine ordering materials for twelve men per day and laying out work with job schedules for each? Just trying to remember what I was doing was a major task in itself.

I noticed that one general foreman watched me each day as I gave my plumbers their assigned work orders. One day, I realized who he was—the plant officer who had interviewed me when I tried getting a steel plant job years ago, when I first got out of the army. I wondered if he remembered me. He was the one who'd told me I had to weight over 160 pounds and that being a plumber's assistant was a white man's job, not a black man's job. He looked angry. Perhaps it was because I'd hired black plumbers in my crew, which never had been done before in the steel plants in Buffalo, New York.

I thought I had better not say anything to him, because union plumbers were not used to having a black man as the boss. That was why I made the union give me six black plumbers—because I knew they'd watch my back on the job, and I'd feel safer. The union complained, saying the black plumbers took jobs away from the white plumbers, but I was the boss, so it was my choice which plumbers to use.

When the job was nearing completion, I finally asked the big-shot general foreman from the steel plant if he remembered me, and he said yes. I asked, "Do you remember what you told me?"

He smiled and said, "Yes, I do. And I've watched you and the black plumbers on this job. I never could have dreamed that blacks could be as good in the trades as white plumbers." Although his comment should have been validating to me, I was amazed that here in Buffalo, people still were so backward in their mind-set. Then his smile broadened and he said, "Will you forgive me? You are the best general plumbing foreman I have ever seen. How can you remember the schedules and how to place your men every day?"

I smiled and said, "That is a trade secret." I felt there was really nothing to forgive because as the foreman, he had just been doing his job. I'd remembered him more because of his high-ranking authority and position than because he'd sent me away, using my weight as an excuse.

Another job was at Niagara Mohawk Corporation, which involved various renovations at sites in and around the city and county. Several jobs were just repairs, but one was installing brand new restrooms. The new restrooms job was an odd-hours type job—we had to work night-shift hours (from 10:00 p.m. until 6:00 a.m.) because we had to turn off the water. Then I had to reschedule my manpower work hours for everyone. Even so, it was such a pleasure working for the major corporations (New York Telephone and Verizon, especially) because they paid very well.

Another major job I did was St. John Towers in Buffalo. My company only did the plumbing contract, but that was my first high-rise building—ten floors. There were a lot of bathrooms and shower rooms—and a lot of climbing steps, because during the construction we could not use the elevator. That job was located deep in the minority community, and I believe they gave me the job because the black plumbers made sure I would get a chance to bid. I think that job made the black plumbers in the union happier because my work crew was two white plumbers and six black plumbers. These black plumbers came from the Project Justice program, and now, as a contractor, I was giving them the proper training that they needed to become plumbers—they often said to me, "This job spoils us."

Working that job gave me a chance to train my first black plumber to be a job foreman on a union construction job in the city of Buffalo, which at that time was a major deal. All the white construction workers noticed this, too. What I did shocked the foundation of the Plumbers Union, because now they all realized that the black man was not inferior; he had been discriminated against for centuries.

After working my union job until 4:00 p.m., I then would do small repair jobs for the elderly in the community. I did most of those jobs for free because these folks could not afford to pay me. When I was in sixth grade, our toilet broke and stayed that way for years because we had no money to pay a plumber. I did the jobs at no cost for the elderly to honor my mother.

CHAPTER 7
SBA 8(A) CONTRACTING, 1985–1990

One of the first jobs I received through the SBA 8(a) program was renovating a hospital's bathrooms and shower rooms. It was made difficult, however, because of the lack of cooperation from the Plumbers Union—at least, at first. This was an out-of-town union that did not know me, so I asked the Plumbers Union from Buffalo to call them and let them know we were brother union members. Unfortunately, we got off on the wrong foot. We argued over my wanting more minority workers because I did not want the entire job done by white plumbers. Also, the VAMC and the SBA both had asked me if I could hire minority workers on the job, and I'd told them, "Yes, I can."

The VAMC and the SBA both indicated that they could have hired a white contractor, who would have refused to hire minority workers—that was what they always did. The government insisted, however, that the union provide me with minority workers because they wanted to hire minorities. Finally, the Plumbers Union supplied me with a black plumber but was still angry with me because I was from out of town. Also, because contractors seldom request minority plumbers, the union initially thought I was messing with them by insisting on minority workers.

I asked one black plumber how long he had been out of work, and he said it had been two years. I then asked, "Are there any more black plumbers in your union who are laid off now?" He told me he knew of some who needed a job. I said, "Have them show up on my job and ask me for a job, and we will call the Plumbers Union from here and pretend they are job searching. Then the union cannot get mad at any of us." The unions are well known for not placing minorities on jobs because in general, minorities are the last hired and the first fired. I, however, was just following my instructions from the government—hiring minorities to work on government contract jobs. The next day, two more minority workers showed up. Although the VAMC and the SBA were happy that minorities were working on the job, most union members thought I was crazy for hiring black workers. Still, I was just doing my hiring honestly, as the government had instructed me, to be fair with minority hiring and training.

What a laugh the plumbers—both whites and minorities—had at lunchtime when they talked about the situation. The Wall Street-money contractors didn't think it was funny, though; they didn't want black workers, but because it was a government job, they couldn't have it their way—and the union was not allowed to picket federal property.

The plumbers on the job, both black and white, were happy to be working and gave me much respect because I had mastered plumbing inspections, job project management, and plumbing estimating.

After my first job with the SBA, I was offered several jobs through the 8(a) program with the navy and with air force and then several more with VAMC hospitals. One job with the air force, however, wasn't a good experience—the air force pays slowly. A white contractor bonded me, and the Wall Street-money contractors loaned me the money to make my weekly payroll. After the job was finished, the contracting officer for the air force refused to make any payments to me. He claimed that I had overcharged the government in excess of the required overhead and profit for air force government contracts, so he was ordering a government audit of my contract price.

This practice is legal, and for this reason, it is important to learn how to keep each job separate and properly indentified in the recordkeeping. If my records had not been in proper order, this contracting officer could have gotten a free job—and it was in a new building; where everything was new and visible to the eye. I knew all the time that the air force contracting officer was dishonest in his dealings with me, because any time a man won't look you in the eye when he speaks, that means he can't even look at himself.

The air force renovation work was easily worth more than $500,000, plus change orders. (Change orders refer to additional approved work added to the original contract price.) The contracting officer had already signed and agreed to the change orders, too. When the inspector general audited the contracts, however, he ordered the air force contracting officer to pay me. Even though I did an honest job in good faith, it took an order from the inspector general in Washington DC to make this biased person in the air force pay me the money I had honestly earned. Many minority contractors face the same type of discrimination.

To my surprise, the contracting officer and engineers at the VAMC told me they were going to give me an office with a telephone in the hospital, because they were making me their in-house contractor; all contracts on SBA and non-SBA jobs would be handled by my company. That would give my business a big boost and possibly mean there would be no more slow periods. The engineers laughed and said that they could use my photographic memory to help them, too!

CHAPTER 8
SBA 8(A) CONTRACTING, 1990–1995

After I won the big SBA 8(a) contract for the VAMC hospitals (worth millions) in the early 1990s, my working relationship with the white contracting firm remained harmonious and strong. When the firm made changes in its corporate officers, it didn't occur to me that it might be serious or any of my concern.

Prior to my bidding the job, I met with government and SBA officials, as there were absolutely no minorities on workforce reports, and they were concerned about the lack of minority worker participation in government jobs. When they asked if I could hire more than one minority on the government jobs and make sure minorities got trained, I answered, "Yes, I can." This job would be more minority-friendly. I would not only make an effort to hire more than one minority worker; I would request the unions to do the same.

Because I was an experienced minority contractor, I knew that many of the white contractors were abusing the real intent of hiring blacks and minorities. They might help a minority contractor get bonding—they called that their minority participation—but they did not hire or establish any training programs for blacks and other minorities. Additionally, these white contractors stuck together because they often were family members, and the family members, together with union members, sat on the boards that governed the makeup of union memberships and training for local jobs. There was not any local minority representation on the boards for any jobs offered in the minority communities, nor were any apprenticeships offered.

Unfortunately, at the time of this writing, this situation remains unchanged. The chances of minorities getting jobs is at a low point due to the discriminatory practices of the hiring boards in construction and of the unions, which are controlled by the Wall Street-money white contractors. This must stop! This must change!

In the early 1990s, the board's selection process for plumbing journeymen and apprenticeship applications each year affected the minority communities' opportunity to learn trades. Most minority communities did not yet have representation, but the unions in this area did not offer solutions to this problem—a problem that tremendously affected the employment of minority youth.

One method could have been to set up a community monitoring agency, with contractors and a union board in a joint effort, to monitor all federally funded jobs. In this way, black and other minority workers could do physical work on jobs with meaningful training, instead of the board's evading or not hiring minorities by using third- and fourth-tier subcontractors, which is outdated—and illegal.

The SBA asked my contracting firm to bid on a West Valley Nuclear Services job because they needed minority participation. They wanted a minority company to hire minority workers, which I thought was really nice. I told them "Yes, I can" when they asked if I could hire young minority folks from Buffalo, New York. The unemployment rate in Buffalo for minority youth was the highest in the country at that time, so it was easy to honor the SBA's request. I simply hired young people who were looking for a job; they all were younger than twenty-five years old. They were highly motivated to learn and train for a nuclear facilities job. West Valley chose my bid, and so I hired the young minorities and made sure they received the proper training for nuclear jobs. These minority workers were outstanding.

All of a sudden, things changed with the white contractors—they tried to hire only white plumbers and ignored me as an equal partner. They also acted innocent when I confronted them about these issues, which were against federal contracting laws. The government had requested that I inform the white contracting firm about these contractors' violations. In fact, all my actions were mandated by the government, such as hiring minority workers from unions and setting up available training programs for minority workers.

The white contractors wanted only one

minority worker hired. I had made arrangements with the union for full cooperation for this job, and I hired five minority workers. The white contractors tried to blackball the one minority worker; fortunately, their attempts were unsuccessful, and this minority worker eventually became our best worker on the job.

At first, the federal officials were angry that I had accused the white contractors of violations, because the white contractors played innocent. The white contractors had their bonding company call me and use bullying tactics to try to get their way—and what they wanted was a dictatorship in running the jobs. I told the bonding company that I had not and would not participate in any violations against the government regulations. The bonding company then threatened that I would never get a bond from them again.

My firm was the prime contractor to the government, with sole responsibility; the white contractors were silent partners, bonding the work for approximately five to six million dollars. Eventually, the federal government stepped in and handled the situation. I only advised the government officials with regard to the white contractors who tried to hire only one minority worker.

The government officials then advised me that my job was to complete the new construction job and hire minorities and provide training opportunities. The government also made special arrangements for me to get bonding from other companies, so all the threats made by the white contractors and their bonding company came to nothing.

The West Valley job people told us that another contract was coming up, and they had to choose between my small company and the white contracting firm, which happened to be my bonding partner. Naturally, my bonding partner knew that their firm was competing against the minority workers. The sons of the Wall Street-money contractors were made officers in the firm, and they as well as others in that firm thought their white workers were superior to the minority workers. Still, their excuse was that they thought it was too dangerous for my workers to do the specialized type of work required for the West Valley job. My minority workers assumed these were just scare tactics, and I stood up for my minority workers.

The next time I visited the West Valley Nuclear Services job, my young minority workers complained about awful and degrading treatment from the white company. I told them, "Now you're in the real world, and because West Valley has just one new contract to award, we have become competitors with the white contracting firm. You have been doing such good work that they really are considering a small, minority company for the big job. This also means if you have questions, please ask me or the West Valley management personnel. Please don't ask anyone at the white company, because we have lost trust in their contractors, who only showed us their greed for a dollar. Please stay away from the white company's personnel, too. Now, I have two jobs where they claim I hired too many black minorities. Both jobs together are in excess of seven million dollars, including change orders."

When the minority workers heard what the white contractor was discriminating against me for obeying the law and hiring minority workers, they won the new contract and showed all that they were not inferior. That was how we won the Contractor of the Year award. (Please see "Special Awards Received" in Appendix A.)

Then, at the VAMC job, the white contracting company tried to incite the white plumbers to not work and to strike against the black plumbers. They said I'd hired too many black plumbers, and whites were losing their jobs because of it.

When I next visited the job, all the plumbers—black, other minorities, and whites—told me that nobody was quitting just because the white contractor said something. As a matter of fact, the plumbers told me that it was the white companies that discriminated against them, not the plumbers union. This was a great victory for a legitimate minority firm and a greater victory for the unions—and a first in western New York. This left the mighty Wall Street-money contractors very upset.

The black and white plumbers all stuck together after they all learned about the trouble with the white contractor. At all my jobs, on three different sites, the black and white plumbers held fast and represented me with courage. And all the jobs were finished after I had my stroke nine months later in 1996. I started my serious rehabilitation in 1996.

Today, all the plumbers in the unions know that when the chairman, president, or any official from a company controls the union membership to satisfy his own interests, it's called discrimination. Plus, this is a direct conflict of interest in union membership rules and regulations. So, now we know that the white contractors—those I referred to as Wall Street-money contractors—were the persons who willfully affected and caused discrimination against minority workers and minority contractors. Their actions and deeds spoke loudly.

CHAPTER 9
FILING OF THE REPORT, 1994-1995

Several years passed after the Wall Street-money contractors and my firm made the agreement with the Small Business Administration that I would be their sole plumbing contractor and do their work as a minority contractor. At that time, the president of the white contracting firm told the SBA that his firm did not want to do any plumbing work, and he would subcontract all his plumbing work to my firm.

The Wall Street-money contractors had several different company officers, with various vice presidents at that time, and all honored this agreement. The working relationship between my firm, the minority community, and the Plumbers Union improved and became better than ever. Then, the chairman of the board promoted his sons to the president and vice president positions, and their corporation became more of a controlled family business.

After that, there was an increase in dissention from the Wall Street-money contractors and the Plumbers Union, as all the racial attitudes toward minorities and minority contracting changed—there was no regard for the laws that were in place, because the sons acted like spoiled children; their father had the money and undisputed control. I tried not paying attention because keeping my nose clean meant staying out of their company business.

But then, the sons tried taking over my minority firm—they would have made me a "black front"—and that's when the real fight against discrimination started. It got so bad that several times, I met with the father, the chairman of the board, alone, so as not to embarrass or disrespect him by speaking with him in front of others, I said, "My position from day one has not changed. I will hire minorities as per my government contract. I run my company

100 percent, and I do my own hiring and firing 100 percent, too. The violations against the government and Veterans Affairs Medical Centers must stop. Tell your sons to back off."

The chairman of the board told me, "My boys are smart and too tough for you. You mean you cannot handle my boys?"

I then told him, "It's not a question of whether they are too tough for me. But I am going to obey the law, and I will hire and train minorities, as per my contract with the federal government."

The chairman then said, "You will be hearing from my bonding company."

Soon after, I did hear from his bonding company. The company told me to let him have his way or they would never bond me again, and my days in construction would be over.

I told the bonding company, "I do not appreciate being forced into committing unlawful acts against the government, and I will hire minorities, too."

After talking to the chairman of the board, I knew that because I stood my ground, I was in terrible trouble. Many black men have fought discrimination, and many have died.

This act of discrimination toward me and the minority community made me turn in a report I had written on the mighty Wall Street-money contractors' violations. If the situation had been reversed and I had committed similar acts, the Veterans Affairs Medical Center and the Small Business Administration would have jailed me right away. I also knew that I would have to file another report, because most black men get killed when they think racist people will not act against them, but they do. I turned in a report to the SBA office in Buffalo

I also requested of the SBA that they prosecute the white contractors for discrimination, if anything should happen to me, because they would be at fault and responsible—they were cold-blooded racists. My SBA representative agreed with me, and I also promised the SBA another report when the job was finished.

The SBA knew that this discrimination battle was intense because the white contractors had the monetary resources to crush me and my minority workers.

I still wonder why, and how, the white contractors continued to get government contracts when they were clearly in violation by refusing to hire more than one minority worker. Their success continues today, even though they rip off the government and the minority communities by not complying with the law.

CHAPTER 10
THE STROKE, 1995

I was getting ready to go to my trade school training for air conditioning and refrigeration and was studying for a test. All of a sudden, something like a lightning bolt hit my head. It seemed to go through my head to my toes, and I dropped to the floor in intense pain. Everything was dark, and my throat filled with blood. Then my vision partially returned, and I could see as if through half a window. My head was lying in a pool of vomit and blood. I tried to lift my head so I wouldn't drown in it, but I could not lift or even turn my head.

Soon after, my wife came home and called for an ambulance. I remember their moving me, but my vision still seemed like

I was looking through half a window—maybe one eye was not working. Suddenly, my head started bursting again, like another lightning bolt was coming. I was afraid, because I now knew what that felt like, and I would do anything to get out of the way. After I got in the ambulance, everything went dark. I remained in a coma for about thirty days.

When I came out of the coma, I realized my head was clamped in a vise-like object. I later learned this was because the doctors had to cut my scalp in half to remove the blood clot. I could hear, and I pretended to comprehend what people said to me, but I could not. I could not move, and I could not talk, and I was confused—I thought that we had lost the world war to Russia, and the Russians had captured me, because at times the speech I heard seemed to be Russian, and I could not understand. I did not know I was paralyzed. I thought the "Russians" had me in chains, as a prisoner of war. So I made up my mind to give only my rank and serial number—but I could not speak either. I thought, *How can I escape from this prison hospital when I am locked and chained in bed?*

Then the doctors told me that I had

several blood clots in my legs that threatened to reach my heart, and if a blood clot were to reach my heart, it would kill me. I still could not talk; I could only listen. They then told me that they had to talk to my wife to get permission to perform an operation, which I later learned was an operation to amputate my legs. I had no say in this decision because I was 99 percent paralyzed, so by law, my wife had the authority to approve it. In the operating room, the doctors saw me profusely crying, as if I was crying for help. This shocked them, as they'd thought I was 100 percent paralyzed! The doctors realized that although I couldn't talk, I could still comprehend. One doctor said, "Sherman, blink your eyes once for yes and twice for no. Do you want your legs cut off?" I blinked my eyes twice, and the doctors stopped the operation. Instead of amputating my legs, they inserted something called a "green filter" in me to dissolve the blood clots.

The doctors later told my wife that it was against the law for them to give me an operation that I did not want. My wife said, "Doctors, he is 100 percent paralyzed. He can't talk to you." But the doctors told her that as long as I could blink my eyes to communicate, I could relay my wishes—and that would stand up in court, too.

After my doctors talked to my wife, they came in to see me and said, "Mr. Turner, you must have some money and success, but in your condition, you will surely will lose your wife. Therefore, it's our job to inform you that you had better watch yourself and not put your trust in your wife." I blinked one time.

Today, I am no longer with my wife, but with a blink of the eye, I saved my legs. God is powerful, because with a blink of the eye things can change. My rehabilitation started with a blink of the eye.

I was in the hospital for six months after my massive stroke. When I was released, I was in a wheelchair but my verbal communication and comprehension skills still were very limited. My job foremen pushed me in the wheelchair so that I could see that jobs were completed and how well everyone worked together, including minority males and females, too. At the VAMC hospital, all plumbers proudly said they enjoyed working together with blacks and whites.

Visiting the VAMC Hospital and seeing

all the VAMC officials, plumbers, and minorities made me cry as I sat in the wheelchair, because we all will remember that job as our coming together for the first time against white contractors whose self-interest was not in the true interest of the SBA or the government's minority contracting programs.

Everybody wished me success in my rehabilitation—I would need to relearn how to walk, talk, and gain comprehension abilities. I felt very sad that I could not thank everyone properly because I was disabled, but I think my tears told them how much I appreciated their completing an excellent job. All my jobs were finished and won special SBA awards, because all the workers—blacks, whites, men, and women—worked excellently together. It made me very proud to see that everyone showed strong courage.

CHAPTER 11
MY FINAL THOUGHTS, 2010

After I successfully completed my rehabilitation, I found out that the white contractors had spoken against me to the unions and suppliers while I was incapacitated. People will always believe what they want, but it's time for me to set the record straight. The government caught the white contractors and called them on their violations but said that it was my responsibility to ensure that they stop their illegal practices. Yes, the government made me, the prime contractor, responsible for my subcontractors' actions in issues of regulation violations. Because of that situation, I would like to request that President Obama create a law to strengthen the existing laws, so the government can deal directly with unscrupulous white contractors when they violate government regulations.

I hope you have gained insight into minority contracting from reading this book. As I mentioned, for me, it was a choice of going into business or going on welfare.

When the SBA and the government asked me if I could hire and train minority workers, I said, "Yes, I can," and that is what I did. All jobs were successfully completed. The government and the SBA can be proud of their minority contractor programs.

Thank you for choosing to read my book.

APPENDIX A

SPECIAL AWARDS RECEIVED—VARIOUS SBA CONTRACTOR AWARDS

- Small Business Award for Western New York (1991)
- Dr. Martin Luther King Community Service Award (1992)
- Secretary's Award for 8(a) and Veteran-Owned Small Business: Department of Veterans Affairs (1994)
- Contractor of the Year: West Valley Small Business (1995)
- SBA Certificate of Commendation for Successful Completion of the 8(a) Business Development Program: Nine-Year Sponsored Government Contracting Program (1997)

APPENDIX B

CONTRACTING REQUIREMENTS THAT SHOULD BE IN PLACE

In every government contract, especially SBA 8(a) contracts, there should be a requirement for minority contractors to hire minority workers. This is an important issue that should be addressed. I think that we minority contractors should be required to hire and train minority workers when working on government contract jobs.

Many minority contractors have become big and powerful, like the white contractors, complete with bonding and plenty of cash flow. To me, it makes sense to enforce new mandates on future contracts that require all contractors to comply or risk losing the contract if not hiring minority workers as required. I feel as though the minority community has suffered enough through the abusive practice of not hiring and not training minority workers.

APPENDIX C

SPECIAL REQUEST—RE-CERTIFICATIONS

I would like to suggest that the minority re-certification process should occur yearly or every two years and eliminate the lifetime certification. I believe this is needed until the abuse in minority contracting is stopped. There is no way to move forward and increase the number of minority workers hired until we get rid of the "black fronts" operating to benefit the non-hiring of minorities so often used by white contractors.

Then, after all the abuse has ended, we can increase minority contractors' certifications to five-year terms but without lifetime certification.

PART TWO

THE SHERMAN TURNER STORY
PART II

CHAPTER 1

THE STROKE, 1995

The Veterans Affairs Medical Center (VAMC) selected my firm to be the government's plumbing and heating contractor because my firm hires minorities and also provides training opportunities. At the government's request, I agreed to be their air conditioning and refrigeration contractor, too, as part of my personal commitment toward helping the Small Business Administration (SBA) attain success in its goal of hiring and training minorities in local communities. Because I was given this additional government contract and responsibility for the government's air conditioning and refrigeration, I decided to go to night school to further my education.

As I write this, I am happily married to a wonderful woman. My wife encourages me to study hard and do my homework dutifully. But I was not always so fortunate.

In 1995, when I was married, I was getting ready to go to my trade school training for air conditioning and refrigeration. All of a sudden, it felt as though a lightning bolt had hit my head and went through my head to my toes. I dropped to the floor in intense pain, and everything was dark. My throat was plugged, and I could taste blood. Then my vision partially returned and I could see, as if through half a window, that I was lying in a pool of blood and vomit. I tried to lift my head so I wouldn't drown in my own vomit and blood, but I couldn't turn or lift it.

When my wife came home, she found me lying on the floor and called the ambulance. I remember the paramedics moving me, but my sight was still limited—maybe one eye was not working. Suddenly, my head started buzzing again, like another lightning bolt was coming, and I was afraid, because I knew then what it felt like. After I got in the ambulance, everything went dark again ... until I came out of the coma, about thirty days later.

I was 99 percent paralyzed when I regained consciousness, and I had to learn to walk and talk again, as well as regain my memory. It gives me great inspiration when I think back on my experience, because when I fell into darkness, I started praying and begging God to remember me. I was so scared and afraid to die. Then, suddenly, there was a speck of light shining so brightly, like a star floating toward me but about ten miles away and within a tunnel. The star became as bright as the sun, so bright that I couldn't keep my eyes open. When I woke up, I knew that God had answered me and had given to me another chance to live and to do the right thing in my life.

The doctors had me in a head clamp, which is a vise-like device, because they'd had to cut my skull in half to remove the blood clot. I could hear, but I could only pretend to comprehend what they were saying. I thought that we had lost the world war to Russia and that the Russians had captured me, because at times, the doctors seemed to be speaking Russian.

I did not know that I was paralyzed. I

thought the "Russians" had me in chains, as a prisoner of war. So I made my mind up to give only my rank and serial number—but I couldn't speak either. I thought, *How can I escape this prison hospital when I am locked and chained in the bed?*

It was several weeks later, after I was able to explain my confused thoughts, that the doctors would tease me and laughingly say, "The Russians are coming, Sherman!" When I first regained consciousness, however, I was able to comprehend their words well enough to understand that I had several blood clots in my legs that were threatening to go toward my heart. If a clot reached my heart, they told me, I would die. Then the doctors excused themselves, saying they needed to talk to my wife about a permission issue.

The next day, I learned that I was to have my legs amputated because my wife had given her permission for the operation. I could not voice my opinion because I was 99 percent paralyzed and couldn't communicate. Therefore, by law, my wife had the authority to make the decision.

In the operating room, the doctors saw me crying profusely, as if crying for help. They realized then that I understood what was about to happen. One doctor said, "Sherman, blink your eyes once for yes and twice for no." Then he said to the other doctors, "Sherman can comprehend what I'm saying." The doctor asked me, "Do you want your legs cut off?"

I blinked my eyes twice for no, and the doctors stopped the operation. Instead, they put in a new device called a Greenfield filter to dissolve the blood clots.

The doctors told my wife that it was against the law for them to perform an operation on me that I did not want or approve. She said, "He is 100 percent paralyzed; he can't talk to you."

They told her that as long as I could blink my eyes, I was communicating. This would stand up in court.

After the doctors talked to my wife, they came in to see me. One said, "Mr. Turner, it seems you have some money and are successful, but in your condition, you will surely lose your wife. It's our job to tell you that you had better watch yourself and not put trust in your wife."

I blinked my eyes one time for yes.

That day, I knew I was going to lose my wife, but with the blink of an eye, I'd saved my legs. Also, as my tears flowed profusely,

I knew, deep in my heart, that God had plans to use a 99-percent–crippled man like me at some point in the future.

CHAPTER 2
IN THE HOSPITAL: PARALYZED, 1995–1996

It took some time before I realized that I was truly paralyzed and couldn't move or talk. I could hear okay, though, so I was eager to listen and find out what had happened to me. By this time, I'd noticed that I was strapped down in bed like a prisoner of war. I wondered what had happened to me because everybody acted as if they didn't know me anymore.

The previous night, deacons, ministers, and church mothers from various churches had been praying at my bedside. I thought I was speaking to them, and I tried to tell them, "Yes, I am here and okay," but everyone kept ignoring me. I wondered why—I didn't realize that I hadn't actually spoken to them.

I looked forward to visiting hours; I was eager to hear what was going on, as well as why everyone seemed to be ignoring me. When my wife came to my hospital room, with her beautiful (and, I later learned, deceitful) smile, she saw that I'd come out of my coma, and she was shocked. She said that the previous night, everybody from the church had circled my bed and said prayers for my recovery.

I started shedding tears because I knew then that I had been with them in spirit only—that's why everybody had seemed to ignore me. At least I knew that I was not crazy. My wife wiped away the tears that fell from my eyes, and then she told me that the doctors didn't want me to move. That was why I was strapped down and had a head clamp, which was holding my skull together.

She told me that the main reason why the church deacons had come from many different churches to my bedside was that I needed prayers because I'd been in a coma for almost thirty days. The doctors had said that the longer I was in a coma, the less chance I'd have to recover. The doctors said my time was running out, so people had come to pray for their plumber. Yes, I said to myself, I do remember asking the dear Lord to please remember me! And begging, too!

I was happy when my sweet sister Denise came to see me. I had never really seen my sister cry for me until that day; she could not stop crying. I guess she was really shaken because I had never been seriously ill before. Denise said she was very sorry for always mistreating me when we were younger. (She hadn't, of course. Sisters will be sisters.) She smiled as she remembered our childhood and said that she always liked the way she could get my brothers and me to give her whatever she wanted. I just smiled and thought, I always knew you were greedy, baby. I just feel so sorry for your husband.

My little brother visited me regularly, too, until I saw that he was ill. Then it was better that he try to take better care of himself. Almost all my relatives came to see me, but my communication level was not there—I still could not speak. Many times, I would miss my relatives the most just after they left the hospital. That was when I felt like I was in prison, with no freedom. The nurses would come to my room every four hours to unstrap me and turn me over so that I would not get bedsores; then they'd re-strap me so I couldn't move.

After three months, the doctors removed the head clamp because my skull had successfully healed. When they removed the head clamp, the nurse brought me a mirror, and for the first time in nearly four months, I saw my face. I felt that I looked so ugly, I asked, "Is that me?" I could not believe that I had so many bumps and knots on my head and that my head was now elongated from wearing the head clamp for such a long time.

The nurses informed me that my "boot camp" would start the next day. I would be trained in how to wash myself, get dressed, and mount the wheelchair. I thought, How can I do all those things when I have only one good hand and one good leg? We started at 7:00 a.m. and trained until 4:00 p.m. five days a week. I also learned how to use the wheelchair correctly.

There were many days when the training seemed harder and harder. I still could not believe that God would put me in this type of situation. I was not a hard-core sinner; I was one of the good guys and always had a good heart. Why would God make me a cripple, a useless person with no future? Sometimes after hospital visitation hours were over, I would sit in my wheelchair, look out the window at the world where I used to live, and start crying. A crippled man's life was a lonely life of sadness. It wasn't right.

It was about this time that my wife started showing signs that she really didn't love me, and I wondered what would happen when I was released from the hospital. Who would help me? Who would take care of me? I was scared to death of my future. Who would protect me? I was in a tailspin.

While still in the hospital, I learned how to mount my wheelchair from the bed and wash myself properly, and that helped me feel more independent again. Then I began working with a speech therapist at the sound and speech clinic. After that, there were grueling exercises to help me learn to walk again. The clinic therapist would decide whether I would do best with braces, a cane, or maybe just a wheelchair when I went home.

All the therapy was difficult for me. Before my stroke, I had weighed 230 pounds; now, I had lost nearly one hundred pounds and was just too weak to benefit from the therapy program. Plus, the drugs I was on made me sleep every two or three hours.

Soon, however, I was able to say a few words, but my speech and pronunciations were terrible and made me feel ashamed of myself. Because of this, I would still pretend I could not speak. During that time in the hospital, some people on the same floor with me were walking and talking and seemed to be doing well. Sometimes, however, when I would try to visit them the next day, their beds would be rolled up and empty. When I asked for them, the nurses would sadly say that the person had passed away. I eventually stopped inquiring because in general, I thought that I was the sickest of all persons in the hospital.

That was when I made up my mind that once I left the hospital, the real battle would begin. I would learn how to walk and talk if it killed me. They could make me take medications that made me fall asleep while I was in the hospital, but once I got home, everything would change. The doctors said I would never walk again, but they never asked me if I thought that I would walk again, nor did they try to help me—other than sending me to rehabilitation therapy—because they had hundreds of other patients. I came to realize that they acted concerned because it was their job.

CHAPTER 3
LEFT THE HOSPITAL, 1996

My health insurance coverage for rehabilitation training was very limited. I was told this was mainly because if I did not regain functions like walking and talking or regaining my memory after a certain amount of time, it was more than likely a hopeless situation. I remembered how my wife had wanted my legs amputated and how the doctors had told me that for most men in my condition, their wives left them within one or two months. Thinking about such a situation made me very nervous, because at that moment in my life, I needed someone to be in my corner and to care for me.

At that point I didn't spend much time thinking about my marriage because the life-saving drugs the doctors had put me on caused me to sleep every two or three hours, or so it seemed. Also, I had a cloth rag tied around my neck to catch saliva, because my jaw muscles were too weak for me to close my mouth. When I finally left the hospital, solely dependent on my good wife, I was in a wheelchair and had developed a dependency to my medication—the doctors had made me a drug addict. I was definitely not in control of my own destiny. Only God knew what the future held for me.

When I was released from the hospital and returned to my own house, my wife, Patricia, helped me up the stairs to the bedroom, helped me get in bed, and then left me alone. She chose to sleep in the newly fixed-up lower bedroom to get away from me. Many nights, I needed something but I could not talk yet, so I couldn't call to her. She'd left me a bell to ring, but she would always say she had not heard it. I felt as though I was not wanted and that I was now a burden to my own wife.

Before I had the stroke, my wife and I would go to church together; after the stroke, my father-in-law and mother-in-law, who treated me as their own son, looked out for me and picked me up to take me to church. This reinforced my belief that God would heal me in God's own time. My in-laws treated me better than my own wife. I was so thankful to feel love from such nice, honest Christians in my hour of need. I could sense that my marriage was in trouble, but I was still defenseless because I was paralyzed, in a wheelchair, and not able to talk. My mother-in-law and father-in-law did things for me, because their daughter would be busy with her friends. They would look at me and feel sorry and would try to make up for my wife's not helping me. I realized that their hearts were pure but that their daughter would soon leave me because I was a cripple and needed help.

Whenever my wife was in the same room with me, I could feel the strong vibes of hatefulness toward me coming from her. She would often yell at me, saying, "How did I get stuck with you?" It was lucky that I could not talk yet and didn't try, because I knew my wife would have a good laugh at my trying to talk. I was in a sad situation and needed support and encouragement. I thought, *At least I know now that my own wife gives me no support or encouragement to learn how to walk and talk.*

CHAPTER 4
WIFE LEFT ME, 1996

Within a month after my release from the hospital, my worst fears came true; my wife left me and disappeared. I was still in a wheelchair. For the first few days I cried profusely, hoping she would come back to help me. After three days, I knew she wasn't coming back to me. My main concern was how I would eat—there was no one to help me. My brothers had all passed away, and my sister was too ill to help. I was forced to try to walk on my own. On my first attempt, I fell to the floor, and there I stayed, trying to get up for what seemed like several hours. Finally, I was able to crawl. Then I just keep practicing until I felt brave enough to call a taxi.

I took a taxi to the bank because I needed money to buy food, but once I got there, I found out that my wife had depleted my bank account eight months earlier— the day I had my stroke, October 15, 1995. She had taken all my money before she came to the hospital to see me. Now I knew why my wife wanted my legs cut off—so I could never catch her when I tried to fi nd out what had happened to all my money. I said to myself, *No wonder she has left me! I had over $100,000 in that account.* So then I checked my other account; she had taken money out, but it wasn't empty.

Although I was shocked that she would take this money, I wasn't entirely surprised. Prior to my going to the bank, she had come to the house to get all her clothes. As I sat in my wheelchair, she told me, "I took all your company's funds and your personal money, too. Catch me if you can." She started laughing and dancing. I was in my wheelchair, crying, and she knew that there was nothing I could do. The more I cried, the more she laughed. Then she said that she and her boyfriend had bought a house with my savings, and there was nothing I could do because her attorney knew the judge, and the judge was already bought and paid. I continued to cry in my wheelchair as she crazily laughed at how helpless I was.

In the days that followed, when I would go to the grocery, I would get very tired. I could not stand for more than five or ten minutes, so it took me at least four hours to shop. And then the lines would be long, so I'd have to wait, and my legs would give out, too. That was the main reason why I started going to the YMCA— because I needed help and additional training to be able to take care of myself. In reality, things happened so fast that I had no time to cry because my needs were immediate. I would find myself praying that I would be able to stand in line; praying to do everything that a normal person does in everyday life.

I called the hospital and made arrangements to take driving lessons as a disabled person. The stroke left me fearful, and it did take time to overcome my fears. Thank God that I remembered that the hospital had programs available for disabled persons. I was afraid of driving, but I had no choice. After taking lessons and knowing which equipment I needed to drive safely, I felt better. I could proceed with my training and rehabilitation at the YMCA. The question was whether they would be willing to take me, because they might not have anyone who could provide the type of help I needed.

Soon afterward, I was in court, using a wheelchair and cane as I faced the judge. My wife's lawyer was the only one allowed to speak. Her lawyer was white and easily took advantage of me—this happens quite often to black men, but it was especially so for me, because I was sick and had lost my memory. I was completely defenseless. The judge refused to look at my medical records to see that I really didn't understand my surroundings because of my stroke. The judge must have been paid of plenty, just as my wife had indicated. The judge said his caseload was too demanding for him to look at my individual case. Then my wife's lawyer laughed at me!

I was at an emotional low. The only thing that compared to my feeling that low was the time when I recognized that the white contracting firm with which I was in partnership was trying to make me a "black front" company for them. A black front company is a firm that seems to be—but is not—controlled by its minority owner.

Fighting for my rights in my company caused me to have the massive stroke that almost took my life. At that time, I could not remember the finer details of my stroke. The doctors told me that it would take time for my brain to create a "patch," meaning that the brain would repair itself, and sometimes it never happened. Although I had no guarantee that my brain would create a patch, I knew I had something to pray for.

All I could do was pray and wait on the Lord to lead me on the right path. I remember waking up from the coma and not wanting to return to consciousness, but I guess it was not my decision to make; it was God's, so I had to start accepting it.

In my prayers, I would promise God that if he allowed me the blessing of learning to walk with a cane, I would not seek revenge on my heartless, money-loving wife. I asked God to give unto her what she deserved. Then I was free to concentrate on rehabilitation and recovery. I couldn't have any distractions if I wanted to succeed.

CHAPTER 5
YMCA'S REHABILITATION, 1996–2004

I didn't know where to start in my efforts to regain my memory. I was too proud to admit to anyone that I'd lost my memory, because I felt they would treat me as if I was retarded—I felt that the stroke had left me retarded. The best medicine for me was to admit the embarrassing truth: I was mentally disabled. My real work would now begin.

Soon after I started rehabilitation therapy, I was able to walk and stand for five to ten minutes. I started going to both YMCAs in the city of Buffalo. The medications that my doctors gave me, however, were not helpful because they made me drowsy and weak. When I asked my doctors what was wrong with me, they said that they'd prescribed the medications because the pain would be too great for me otherwise. I knew that I had to stop taking those medications on my own if I wanted to get better and do more rehabilitation training.

At the YMCA on William Street in Buffalo, I saw my good friend Heide at the swimming class—Heide had been a champion swimmer when he was younger.

He asked me, "What are you doing here? What happened to you?"

"I had a stroke, and now I'm paralyzed," I told him.

"That's too bad," Heide said.

The doctors had said that in order to regain my balance and learn how to walk again, I should walk in the water, while wearing a life preserver. Because I was still partially paralyzed, I said that I would feel much safer if Heide were watching over me and giving me instructions. Heide agreed to watch over me on a regular basis and made me promise to come back. I guess he could see that I was struggling badly because sometimes I did let my disabilities get the better of me. Soon, everyone at the YMCA pitched in, encouraging me to walk again in the water.

My legs started getting straighter because the water made them stronger, and I noticed great improvements. Today, I understand more about patience because when I was crippled, I more or less had to wait on God to fi x me, both physically and mentally.

It wasn't long before different friends at the YMCA— much to my surprise—started helping me, just because they were nice. Sometimes the ladies would even offer, not wanting the men to outdo them; soon, everybody started spoiling me.

When I trained at the YMCA on East Ferry Street in Buffalo, I met my good friend Vernon, who would help me like I was his little brother. I was trying to regain and rebuild my balance, with strength and coordination, but at the time, I only had the use of one hand and one leg. Vernon would teach me how to use my cane for exercises and how to get better balance. He set me up with special classes when he was available. Sometimes on Saturday, Vernon would make special arrangements at the YMCA swimming pool for me so that I could learn to walk in the water; this helped build up my leg muscles and improved my balance.

Vernon always insisted that I take my time doing my exercises and do sets of five only, for starters. Then, with his permission, I could try sets of ten. As I've mentioned, my medication made me sleepy, so I just stopped taking it—without the doctors' permission. I felt I didn't need it because they had me on those drugs to prevent me from committing suicide. I remember when Vernon said, "Do I make you so tired that you have to sleep when you are supposed to be working out?" Then I told him, "It's the medicine and not the exercising that makes me tired." This gave me a reason to stop taking the medication.

Because I was still partially paralyzed, I thought it would be best if I could learn some self-defense tactics to protect myself—and because I thought it would be fun to learn. I knew a Golden Gloves champion boxer at the YMCA, so I asked him if he could teach me some self-defense moves.

He agreed, and we both started dancing around the ring, to the left, then to the right, then backwards. Then he did the "Ali shuffle," and with a boom-bam-boom, I went down for the count. I looked at him with disgust and said, "Why did you hit a cripple that way?" He told me, "You left yourself open." I said, "I wasn't open; that's my paralyzed side, which is defenseless." We both started laughing because I soon learned there was no self-defense for a crippled person like me.

CHAPTER 6
MY GUARDIAN ANGEL, 1996–2010

My friend, whom I'll call William in this book, was my "guardian angel." He would go to the YMCA with me, just to make sure I didn't fall down and hurt myself. (I won't use his real name here because if his identity became known and the Wall Street-money contractors found out that he'd helped me, they would blackball him in the construction trades in western New York State. This has always been the treatment for blacks and other minorities because of the unscrupulous control and power of the Wall Street-money contractors in this area.)

William's uncle, a black master plumber and another of my friends, had made arrangements for William to help me while I was hanging around his plumbing offices, retraining myself, before I went to the YMCA. At that time, William was twenty-two years old and had finished college with an engineering and architectural degree. In my spare time, I would give William lessons in plumbing estimating. I would also do free estimates for his uncle, the master plumber, because he allowed me to retrain myself at his offices.

I also learned how to use the newest computer programs for estimating and learned how to type sixty words per minute with one hand. It took some time, but I also learned how to write left-handed. Eventually, I noticed that my skull did not hurt as much as it had before. Could this mean my brain was healing and maybe my memory would return soon? This made me sign up for classes at the Buffalo Board of Education's Up-Skill Academy. I discovered that the Up-Skill Academy would do an evaluation to determine my comprehension level after I had the stroke and work to bring my levels back to where they were before the stroke.

My guardian angel agreed to teach me architecture and engineering prior to his attending job construction meetings. Then I would look at the drawings and specifications before giving him my answers to his questions. Soon, William said that he saw my improvement with the blueprints, because my comprehension level and retention abilities were vastly improving. Previously, when I tried to concentrate, the place on my skull where they'd operated would really hurt, but by this time, my head hurt very little.

I could tell that William was happy for me, because his encouragement was genuine. I had improved and now it was time to move up to the academics, maybe trying to learn a foreign language, to simulate my brain cells for further improvement, if possible. The doctors had advised me to take it slow, because pushing too hard usually resulted in the patient's giving up, and that was something I didn't want to do.

CHAPTER 7

UP-SKILLS ACADEMY, 1997–2002

After I was able to walk at least twenty feet, I start working on my memory, speech, and comprehension skills. The Buffalo Board of Education had an excellent program for people who are slow or mentally challenged. I had taken an aptitude test to determine in which grade level the school should place me. I was not surprised that they started me at the third-grade level, but I did feel ashamed that I could not do better on the tests.

I was in the third grade again, and this time I noticed that I was the dumbest student in the class. Naturally, I would try hard, but nothing would stick in my brain. One fellow student picked on me because I couldn't retain anything. He got on my nerves, saying, "Nobody is that dumb. I know you can remember." Then I would remember what my doctors told me—to keep trying, because after my brain surgery, everything had to regroup. My new brain cells had to grow and replace the dead brain cells. "Each person is different and unique," they'd said. "Don't give up."

I continued my training at the YMCA while also going to school, and after six months, I was promoted to the sixth grade. One day in class, the instructor gave us a square-root problem to do. I shut my eyes, and I could see the problem in bright lights—and the answer, too. My instructor looked at me and said, "What's your answer?" I told her my answer and said, "My memory came back to me." My instructor was so happy, and I was elated, too!

I continued classes until the year ended in June. I thanked everyone for helping me in class and with all my studies; I wished that God would bless everybody, too. Sometimes I felt that God was watching over me with special care. I did understand that I'd had only a partial memory return but it was a good beginning. My instructor advised me to take college classes in night school to enhance my memory and allow my brain cells to do a complete rebuilding and healing.

By this time I had retrained myself to type and write, and because I had the use of only my left hand, I became left-handed. I decided to retrain during the summer at the computer school. Then maybe I would be able to keep up with those college students; maybe college would be a nice experience for me, too. I had taken short courses to upgrade my knowledge of technical plumbing mechanics but I had never tried learning a foreign language. My doctors suggested, however, that learning a new language might help me. They said that the challenge would encourage the brain cells to expand and to replace the dead brain cells with new brain cells, especially on the damaged side of the brain. I decided to try because I didn't have anything to lose, even though I might be embarrassed in college if those young students laughed at me, the old crippled man.

During the school year of 2001, I noticed for the first time that my memory was returning, little by little, because every now and then I would get sudden flashbacks of how my workers were being discriminated against by the unions and on the jobs. This made me eager to know who the person or persons were who discriminated against my people.

CHAPTER 8
BUFFALO STATE COLLEGE, 2002–2004

It was a turning point in my life when I realized that I was brave enough to face the whole world while I was crippled, paralyzed, and with no memory, yet I was going to a place of higher education. I believed in my doctors, and I also was getting the feeling that learning my roots of my mother tongue would benefit me. Afraid or not, I knew that I must do this.

I eventually found out something that was so interesting, it blew my mind. I found out that what I'd learned in school was all lies, and this was my only opportunity to learn the truth about my heritage and Africa. I would do this for Ma; she would be proud that I wanted to learn about our people and our mother country.

Buffalo State College waived my fees, so the classes were free. I attended classes there from 2002 through 2004, learning Kiswahili, also known as Swahili, an African language spoken primarily in eastern and central Africa. On my very first day in class, I was embarrassed because I was a sixty-three-year-old man, and the other students were so young and very sharp-witted. As the professor explained things, the younger students seem to catch on quickly, but I was left wondering what was going on. I could not learn anything because they were going too fast for me. I despaired that I wouldn't be able to learn a new language.

During the break time the professor asked me, "How are you doing, Bwana Turner?" I told him, "I lost my memory, and my doctors told me I should try learning another language, because that stimulates and helps build brain cells. It might help me with my memory loss." The professor said "Just keep trying, even when you get disappointed. Just like it took years to learn to walk again, it takes time to learn a completely new language."

After about three months of learning Swahili, I started to comprehend it, and my brain was able to evaluate and process the information. Somehow, that process brought flashbacks of different incidents; it jogged my memory of the discrimination problems that I'd faced prior to having my stroke. When I went to sleep at night, I would try to dream and make the flashbacks come so that I could learn with greater understanding. In the morning when I would awake, I felt disappointed because I wanted to understand and know the mystery of this discrimination. I knew for sure that there had been acts of discrimination against me and the minority workers, but I could not remember the specific details.

Near the end of my first year of learning Swahili, the professor asked the class who was coming back for the second year. Naturally, I raised my hand, not because my final grade was an A but because I liked learning my mother tongue. America was the land of the American Indians, which was my paternal grandfather's ancestry. My maternal ancestry, however, was rooted in Africa. Learning about my ancestry made me want to visit Africa. What I learned in college was new to me because all the white schools I'd attended as a child had taught me that Africans were dumb and ignorant people.

My second year at Buffalo State College was the most interesting, because I finally got confidence in myself with my studies and my planning methods. I earned an A on every test, and I loved the language—Swahili speaks directly to the point. The classes were fantastic because the professor loved his motherland; and his pride was obvious in his speech. This was the first time I ever experienced a real teacher doing the job of teaching—it had not been like this in the ghetto schools of my childhood. I will always be grateful to the wonderful professor for giving me the chance I needed to develop comprehension and patience.

Then in 2004, I was practicing plumbing estimating on the computers in the small space that William's uncle had set up for me in the basement of his plumbing shop. I saw a white contractor's sons arrive and rush out from their trucks and cars. I wondered why they were there. I didn't want to see them, so I went back to my small basement space in the plumbing shop. In my flashbacks, I'd been getting strong indications that the root of the acts of discrimination involved these sons.

Now, they used two different entrances to enter the building, and suddenly, they were at my office door. They greeted me and said they'd come to see William's uncle, but then they left without saying anything else. I decided I had to leave the plumbing shop immediately, but I couldn't hurry because I was crippled. One brother caught up with me in the parking lot, but he only said good-bye. My complete memory had not come back at that time, so I did not know what else to say to them. I'd had flashbacks, but I felt I shouldn't prejudge them before all the facts were clear. I would have to wait until my memory came back. I had a strong vibe that the white contractor and his sons were the ones who had shown racial discrimination on government jobs when I was hiring minorities. I still wasn't 100 percent certain, but I'd had brief memories of their actions. It was scary to me that they'd confronted me, especially because I was a defenseless cripple, and I had good reason not to trust them.

At home, I got on my computer to

investigate their business and found out that they no longer had a $10-million-a-year business; now it was $100 million a year—and still growing.

I believed that they were afraid that because they had federal government contracts in the hundreds of millions of dollars, any type of investigation into their discrimination practices would hurt their business. Plus, any type of discrimination case would open an inspection that could show they might owe monetary refunds to the government, minority contractors, and the minority community.

I decided it would be much better for me to go to Kenya to continue my rehabilitation. I needed to get away from these men—I didn't like the way they entered the building and just bullied their way into my space when I was a helpless cripple. Plus, the weather would be better in Kenya during the winter. The winter in Buffalo would bring additional hardships for me, because it was even more difficult for me to walk on the slippery snow and ice. I started making serious plans to go to Kenya for my rehabilitation. Even though I had never traveled before, my doctors encouraged me to make a strong commitment to try harder in Africa if I was willing to travel abroad. It took me at least six months to make the necessary arrangements to travel to Africa, including getting my passport, medical shots, and other security clearances, but finally, I was ready to go

CHAPTER 9
TRIPS TO KENYA, 2004–2006

I did not have a traveling companion for my trip to Africa, and it was hard for me to travel alone. Still, I was determined to go. The flight to Kenya was long—almost twenty-seven hours, with layovers and three planes to catch at three different airports.

When I reached the airport in Nairobi, Kenya, I learned there was a problem with the plumbing—there was a pressure pipe leaking. This meant someone would need to shut down the main valve or section valve and then drain the piping to fix the piping system. All of a sudden, I had a flashback, and parts of my memory started kicking with regard to fixing plumbing system leaks—I had installed and repaired so many for the Veterans Affairs Medical Center (VAMC) as the government's contractor. I saw a man catching the water in a bucket, and I told him they needed to shut off the valve so that the leak could be fixed, instead of closing an entire section of the public airport. Alternately, they could freeze the line on the pressure side and then repair the line on the drainage side. My memory was coming back!

I arrived by taxi in Mombasa, and I was excited and joyous. My Kenyan *mwalimu chango*, or headmaster, greeted me with "*Karibu sana*" (meaning "welcome"). He then introduced me to Bibi Rachel Onzere and her three sons, with whom I would be housed during my training at the Wassermann Language and Rehabilitation Center. Later that afternoon, Bibi Onzere drove me to the Wassermann Center so I could look around, meet my instructor, and learn my schedule. Everyone was nice but the facilities were impoverished, and for the first time I saw that the people of Africa were very poor but seemed happy and had much pride.

The next day Mwalimu Iha (Instructor Iha) started me of in class with a review of the basics in mathematics to find out how the stroke had affected my thinking and my memory. Our class included two other students, but I was the only stroke victim. Mwalimu Iha had a unique way of teaching; he would say, "Are you with me?" or "Are we together?" which made me smile. The other students were a German and an Italian. Iha would speak to me and explain things in English, then speak and explain things in German, and then speak and explain things in Italian. This really amazed me because I had never seen anything so masterfully done. Plus, when an African would come in the room, he would speak Swahili and sometimes their native tongue.

One Saturday, Bibi Onzere invited me see some sights with her three boys, Mike, Timo, and Nelson. I was amazed to fi nd out that Bibi Onzere could speak Swahili, English, French, and two other African languages. Her sons all spoke three or four languages, even though they were just young teenagers. This was really amazing to me. Then I would think how my doctors had said that people who speak other languages really don't have strokes. Very few people in Africa have strokes.

I soon developed a profound respect for the people in this very poor section of Africa. They were poor economically but not in family values and not in their desire to get an education. After I came home from my training and rehabilitation school, Bibi Onzere's boys would drill me on my homework assignments, and then we all would start playing different word and language games. I knew that I couldn't compete against them because they were so smart, but the boys never gave up trying to help me with my memory and rehabilitation.

I believed that soon my complete memory would return. One day, Bibi Onzere told me that her friend wanted to meet me because I was from America. Her friend's name was Bibi Velma Onyango, and she was related to Senator Barack Obama in the United States. I wondered if she would be disappointed because I had never heard of him, maybe because I was from New York.

When I met Bibi Onyango, she was so excited that I might know her Senator Obama that I felt bad telling her that I had never met him. Then we went shopping at

Nakumatt, a Kenyan supermarket chain. Bibi Onyango saw a picture displayed of Senator Obama and said, "There is his picture; that's what he looks like. All the people smile when they see the 'Son of Kenya,' because in Kenya, he is more popular than Kenya's President Kibaki."

I asked her why Obama was called the "Son of Kenya" when he was born in the United States. She answered, "Senator Obama is our symbol of hope, hard work, and education for the children and us poor Kenyans. You know his father was a scholar and spoke many languages, and he is smart, just like his father."

I thanked her for the information and then suggested we visit the gym, because the next day I was to start my workouts on my leg and arm.

At the gym, the trainer, Jako, asked me if I knew the Son of Kenya, Barack Obama. Then he said, "I promise you that if you work hard and train hard, like Barack Obama, you will overcome your limitations and regain your balance to walk again." I thanked him for his wonderful encouragement.

Jako would start of my training and then leave me so that I could do my training workouts alone. I often would get friendly visitors, asking me if I knew their Son of Kenya, Barack Obama, and many people would say that they were his cousins. That is how I got to meet most of my friends in Kenya—the friends of Barack Obama would always be nice enough to help me in the gym.

I took my Kenyan driver's license test in 2006, and I had to report to a high-ranking police officer. The clerk told me to knock on the door and then enter the room, which I did. All of a sudden, a giant of a man, about seven feet tall, shouted at me, "Don't you know how to knock and salute a senior officer?" Then he shouted, "Go back and re-enter properly, soldier!"

I went back out and re-entered, saying, "Sherman Turner reporting, sir," at the very top of my voice, like a real man. The officer smiled and said, "What do you want, soldier?" I said, "I am here for my driver's license test." He asked if I had an American driver's license. I quickly rose to attention and said, "Yes, sir, I am a licensed driver in America."

Then the officer rose and took a deep breath. He asked me, "Who are you voting for in the presidential primary election?

Mrs. Clinton or Obama?"

I smiled because I was wearing my Obama cap. I quickly said, "I am an Obama-for-president man."

The officer asked, "Why do you think Obama is going to win?"

I said, "We keep hope alive, and we are free to vote for the best man without being oppressed."

"Then you're saying Obama will win?"

"Yes, sir!"

The officer stood up, smiling, and I offered to give him my Obama cap. His face lit up like a Christmas tree. He took my Obama cap and said, "When Obama wins, come visit me so we can celebrate together." Then he allowed me to take the driver's test, which I passed.

As I was leaving Kenya on the flight back to the United States, I thought about the time I was in the hospital, and the nurse had left a razor for me so that I could shave. Instead, as I lay in the bed, I tried to cut my wrist. Because I had the use of only one hand, I was not strong enough to break my own skin. The next day when the nurse came back, she laughed and told me that it was God's will for me to survive, so I might as well accept it. Now, I saw it also had been God's will that I should live so that I could visit my motherland.

CHAPTER 10
RACHEL, 2004 AND BEYOND

I met Rachel Onzere when I first came to Kenya. She was the woman who provided me with a place to stay when I attended the Wassermann Language and Rehabilitation Center. My rehabilitation would not have been as successful if not for her prayers and kindness. She always made sure I had the proper travel arrangements so that I could attend my rehabilitation classes, and she took my schedule very seriously. Her sons always looked after me—it was as if I was their child and they were my parents. I owe Rachel and her boys my thanks for making my recovery very successful.

One day in 2004, the water in Rachel's house just stopped flowing. Rachel asked her boys, "Is there any water in the barrels outside?" There wasn't, because it hadn't rained yet. You see, in Africa they catch rainwater for washing, cooking, and drinking because there is a shortage of water in Africa. Then Rachel started crying and saying, "What are we going to do? I cannot afford to call a plumber, and you children need to wash and eat. And we have a guest, too!"

I asked Rachel, "What's the matter? Can I help you?" She looked at me and started crying more. So I asked her boys—Timo, Nelson, and Mike—to explain the plumbing layout to me so I could understand how it was done in Africa. With my good hand, I drew the layout the boys described to me, with water meter, water piping, the main shut-off valve location, plumbing fixtures, and the water tank above the ceiling. I knew then that their water system was a down-feed source.

With the water tank's position above the ceiling, I knew that might mean trouble at that location. Timo said, "The last plumber who came here said we needed a new water tank."

I said, "The other plumber might have been right, but let's do some engineering. If the water tank is empty and has no water in it, we have to find out why."

The boys all agreed to help, and then we saw that the existing galvanized piping was only about five years old and in good shape. That told us that the galvanized piping must be plugged with rust someplace in the system. But where? I said to the boys, "Follow me. We'll start at the main water valve, tapping the galvanized piping to gently rattle loose the rust and let the water pressure wash the blockage clear." It took about ten minutes of tapping the water piping, and then the full water pressure came into all the plumbing piping and fixtures in the house.

Still, the water tank was not receiving water like it should. I asked the boys to climb above the ceiling and draw me a picture of the inside controls of the water tank. Rachel's son Timo was hoping to become an engineer, so I knew he understood mechanics, and he was very smart. Timo handed me his drawing; it looked like an old-time water float valve, like in our American toilets. So I told Timo and the other boys to get their bicycle tire-patching equipment, and we would fi x the water float valve.

Timo and the boys climbed up to the ceiling and into the water tank and started cleaning the water valve parts with sand-cloth and rags. Then I instructed them to coat all the parts with grease—we used Vaseline— after cleaning the parts. Then the boys turned the water back on, and everything worked perfectly fi ne now. Rachel saw that the water was running perfectly fine and that the water pressure was more powerful—it had increased tremendously. She and her boys were all smiles.

Soon afterward, Rachel and I started dating while I continued my rehabilitation training in Kenya, where the warm weather benefited me very much. Spending time with Rachel was also a benefit of being in Kenya, and it became a joyful pleasure to my heart each and every year. She and the boys were always ready to receive me with open arms. From 2004, when I first visited Kenya, until 2010, I took about fourteen trips to Kenya. We still laugh about my many trips—I said that I needed the special rehabilitation training. Rachel and her boys said they all enjoyed our learning together and especially having dinners together.

The culture was very different in Kenya, and one thing I noticed the most was the respect for the mothers and fathers. There also was respect for elders and for the disabled, whether it was someone who had all his limbs or someone who was more severely disabled, with missing or non-functioning limbs. Africans have strong family values and are able to trace their family histories. We African Americans have lost our family history through being slaves; we have lost our roots and that has stunted our growth.

Rachel and I didn't talk a lot about having a serious romantic relationship during our early years together, because we both had experienced bad relationships. But as my health improved, we both realized that we had been together for so long that not being together was out of the question. We planned for her to visit me in America so that we could marry in the United States, which we did on August 22, 2009.

CHAPTER 11
TRYING TO GET A JOB, 2006–2008

From 2006 through 2008, I tried to get a job, part-time or full time, without success. I felt I had been blackballed in the plumbing industry in Buffalo because of the powerful influence that the white contractor had in the area; he was the most powerful and richest Wall Street-money contractor in this area. I had received many hints that no one wanted to go against his wishes by giving me a job. This method is still used today to deny jobs to many minorities in the construction trades in western New York.

I knew that job training would help my

continued rehabilitation efforts, so I decided to seek a job with the SBA emergency disaster services in the flood areas; they had advertised that they needed experienced inspectors. I applied for the job, and when they saw I was crippled, they asked if I could operate a computer and how many words per minute I could type. I told them I could type fifty to sixty words per minute with my one hand. The SBA official then replied, "We'll see how you work out in training first."

This was the first time in about thirty-five years that I'd been given a job as an ordinary worker or laborer that was out of my field of expertise. I received special training for the job because as a computer operator, I had to answer phone calls, and I was supposed to follow a prepared script on all calls. The government monitored the calls to make sure I followed the script.

I thought that memorizing the script would be too tough for me and that I couldn't function at such a high level because I was still partially paralyzed. The other more serious problem I had, though, was that after the stroke, I talked more slowly. This meant that when a person called, he could dominate and control the conversation because I was not quick enough to start talking. I soon found out, however, that I just worried too much—I did very well because the SBA team took their time and trained me well.

The next paid emergency/disaster job training I had was in Fort Worth, Texas, for SBA, for two weeks as a field construction inspector; it was computer training. When I entered the classroom, I knew that everyone wanted to laugh because I heard sarcastic remarks about my walking abilities. They said things in a taunting tone, such as "Can I *help* you?" or "May I be of any assistance?"

As we got to the part of the course that discussed the requirements of field construction inspector, I was the only one there who could read blueprints—that meant that I was the only one qualified because I'd already had construction inspector training. During our field excursions, everybody tried to team up with me so they could get passing grades in this class. Naturally, that gave me a good feeling, and these classes helped me in my rehabilitation also.

It felt good being able to function again; to travel and learn how to take care of myself again. To me, this was very important because after my stroke, few people had the time to help me. Obviously, I am very appreciative of all those wonderful people who did find time in their lives to help me when I needed it.

Each time I got laid off from a temporary job, I would mail my résumé, including my special computer estimating experience, to every plumbing company in the city and suburbs, hoping to get a full- or part-time job. This, however, was hopeless, because the "Wall Street Money Contractors" had all the control in this area, and no one wanted to make them unhappy by hiring a black man.

To keep up with my rehabilitation, I would buy the newest computer plumbing estimating programs. I would study intensely at home and then again try to get a job by mailing my résumé and job experience to various plumbing companies in the area. I still did not have any success. I don't want to seem like a crybaby, but I didn't know why no one would hire me.

I went to my friend, a black master plumber. He couldn't hire me because he had no available jobs, but I showed him my new computer estimating system, how it worked, and its advantages, and he was amazed. I asked him if I could estimate some jobs for him at no charge, just so I could become an expert operator on this new system—it would also help my rehabilitation efforts. He gave me his approval.

I did this volunteer work, as a plumbing estimator, from about 2006 through 2008. My friend's company started winning jobs, and that made me very happy. Using my newest plumbing computer systems was nearly perfect. Plus, the getting the practice helped me become an expert. I could estimate, on average, one-million-dollar jobs, from the ground to completion, within just one or two weeks, which is very good in commercial work. We then only had to wait for the material prices from the material suppliers and the different subcontractors' pricings.

I completed a very advanced one-year computer retraining, using the newest computerized plumbing estimating system, which greatly enhanced profits and increased the chances of winning up to 99 percent of the jobs bid. And I learned AutoCAD 2008 for construction plumbing in night school, too, as this was most used in commercial construction.

CHAPTER 12
VISITS TO KENYA, 2007–2010

I remember wearing my Barack Obama hat to Kenya in 2007, and when I arrived at the airport, I got smiles and salutes from the Kenyans there. I had brought about twenty-four Obama caps with me, so I passed some out to the Kenyans who helped me with my bags as I transferred to my next flight to Mombasa, Kenya. When I arrived in Mombasa, all my friends were waiting. They wanted their caps so they could wear them at the airport. It seemed like a party at the airport, and as we left, wearing our Obama caps, we were calling out, "Yes, we can!" And all the Kenyans in the airport started chanting, "Yes, we can!"

When I trained in the Kenyan gym, my friends would always ask me, "Do you think Barack Obama has a chance of becoming the president of the United States?" And I would always respond with the slogan of the Obama campaign, saying, "*Dio tunaweza*" (Yes, we can). Then the whole gym would join me, saying, "*Dio tunaweza!*" That was the joyous part of my day, because then Kenyans would tell me that they were related to my future president and really make me smile.

One Kenyan said to me that he was my future president's cousin and also was a special rehabilitation instructor. He offered to always be my personal instructor when I came in to the gym. He was really good at his job, because my paralyzed hand and fingers started moving after doing his exercises. This made me feel that if I could stay there for at least one or two years, I would be able to regain movement of my paralyzed hand and fingers.

Everything wasn't positive in Kenya, however, as I saw that the people were oppressed. When the young Kenyans would walk by a picture of Kenya's President Kibaki, they would not acknowledge that the picture was there—they would keep on walking without looking at it. This was a prevalent attitude as I traveled the country. I could feel the tension rising, and the national uprising over the Kenyan elections was not a surprise for me. I had been at the gym among the younger and poorer Kenyans and had a great understanding that the youth would change Kenyan politics, just as Barack Obama, the "Son of Kenya," would in America.

When I went shopping at Nakumatt, I would see the very large picture of Barack Obama hanging in the store, and every time the people walked by, they would look at the picture and smile, proud that he was the "Son of Kenya." This was present all over Kenya. I told my friends, "It might be my old bones aching, but it seems like something strange is in the wind." They agreed with me that the winds had changed, and Kenya was going to see a new way of doing free elections soon.

It was not surprising when I saw bloodshed in the streets between the political parties and the different tribes in Kenya. The uprising was also between the youth of Kenya and the older folks, who did not want change. My friends were the youth in Kenya, even though I was elderly. In the streets at night, in the darkness, I could hear the people chanting the name of the man they hoped would become president: *"Odinga! Odinga! Odinga!"* The young people in Kenya wanted change, and it would come soon.

On August 4, 2010, Kenyans voted on whether to adopt a new constitution. The winds had changed, and constitutional referendum won by nearly 70 percent of the popular vote. President Obama sent his congratulations to the Kenyan people, and Secretary of State Hillary Clinton praised the Kenyans for turning out in large numbers to hold a peaceful vote for constitutional reform and wanting democracy. The passing of this referendum made me so proud of Kenya and my brothers, the African leaders who stood up against racial discrimination, because I am personally against the discrimination of any and all people.

I had decided that I would visit Kenya again if Senator Barack Obama won the US presidential election because I did not want to miss the celebrations with my Kenyan friends. They had stood by me during my rehabilitation years and had given me encouragement. After working hard with my friends in America to encourage folks to get out to vote, as I had promised my good friends in Kenya I would do, I thought I owed myself the treat of seeing my friends celebrate in Kenya—Kenyans know how to celebrate, and they liked our new president more than they liked their own president. When I visited Kenya in 2009 and 2010, there had been mixed reactions over the fixed presidential election of Kibaki. Also, it was a known fact that our president, Barack Obama, wanted Kenya to have free elections. Based on my visits to Kenya, I could tell that President Obama was more popular in Kenya than their own president, Kibaki.

When Kenyans realized I was from America, they always asked me, "How's your president?" and that immediately brought joy and happiness, because we were now together in friendship. The president of Kenya had to recognize the US election of President Obama, and he gave the Kenyans a paid holiday, just to keep the peace—this was a government-paid holiday in Kenya, which was rare.

I went to the barbershop to get a taste of the people's reactions to the Obama victory. Prior to Obama's winning the nomination, most Kenyans did not believe he had a chance against a white lady, especially one who happened to be President Clinton's wife, in a white country like America, which had a history that included slavery. As soon as I walked in, everybody smiled. They knew I was back because I'd told them Obama would win the nomination and the election—our candidate for president was the best man for the job.

Next, I visited Nakumatt. In Kenya, Nakumatt sells things like groceries, building supplies, and furniture, and there are movies theaters, too, in a large complex. My many friends still remembered me, and they all gave me nice smiles because they knew that Barack Obama was now the president. This encouraged all Kenyans not to give up their next elections for *uhuru* (freedom)—they did not have the freedoms that we in the United States enjoy and take for granted. For example, when I was a paperboy, many advertisers would ask me to insert their ads in my daily newspapers, and they would pay me money to do them that favor. I was free to decide for myself whether I would choose to do that. At that time in Kenya, however, people were not free to advertise. People were not free to decide for themselves, because the government controlled the people and everything they did. Kenyans had to get the government's permission to do anything.

It reminded me of the "Jim Crow" days in America, when the white Wall Street-money contractors claimed that blacks were free and equal, but there were no jobs for minorities, because the whites owned and controlled all the companies and the unions. Then I would get stronger flashbacks of how my workers on those jobs were discriminated against, The SBA enforced our rights to work on government-funded jobs, but they couldn't force the other workers to treat us with respect.

CHAPTER 13
COMPLETE MEMORY BACK, 2010

Between 2006 and 2008, there was racial tension and discrimination in Kenya between different political parties and clans because of the upcoming elections. Seeing people get slaughtered or even hurt over these racial and political matters gave me sudden memories of what I had endured while working on jobs for the Small Business Administration 8(a) program, West Valley Nuclear Services (WVNS), and the Veterans Affairs Medical Centers (VAMC) jobs, the stress of which caused my stroke. Just the pressure from one job would have been more than enough, but handling two or three jobs for the government when there was discrimination and fighting a white contractor—that was heavy-weight fighting.

Over the years 2009 and 2010, I did remember more details of how I'd told the white contractors that their people were not complying with the federal mandates in my agreement with the federal SBA 8(a) contract. I had to move my company operations away from their offices, because I wouldn't have any dealings with those who would not hire minorities on the job. It had been suggested by the white contractors' people that I hire only one minority to every ten whites, as that was the census quota—at the time, whites did not outnumber minorities on a 10-to-1 scale in the United States. I refused to comply with that request, which made my moving out of their facilities justifiable. They now wanted a black contractor only as a front.

Then in late 2010, I finally remembered the fi ner details of the discrimination and faulty practices and the times when the VAMC had come to me. They'd said that they knew that the white contractors and their people were planning to commit violations against the government regulations, to get even and mess up the job. I remembered now that the VAMC officials had confided in me several times, and I'd suggested a way in which we could remedy the situation, because I had expertise in construction and inspection, too. It had been an all-out war—the "Wall Street Money Contractors" had decided to make the job harder because they wanted to see minorities fail. It never made sense to me, but that is why discrimination is so wrong. Ultimately, I teamed up with the SBA, the VAMC, and the WVNS so that the minority community and the minority workers would not get the blame for a faulty job.

At least now I could rest, knowing that my memory was intact. I was so thankful to all those wonderful American doctors who suggested that I learn a foreign language. Choosing the language of my ancestry helped me become more thankful to be a US citizen, because in Kenya, I saw again how discrimination destroyed the weak and underprivileged people.

CHAPTER 14
CONCLUSION, 2010

Thank you very much for reading my story. I hope that my experience offered something of value. The valuable lesson that I learned, personally, was that I will never place material wealth above God or seek more worldly wealth. Someday, we all will be judged by our word and not our wealth. I know for sure that when I needed God and called for God to help me, God came to my aid.

Minority workers won and achieved special awards for their work on the West Valley Nuclear Services and Veterans Affairs Medical Centers jobs, showing that minorities have skill in the construction trades. They should be awarded special, targeted jobs so they can avoid having to place their trust in unscrupulous "Wall Street Money Contractors" for bonding purposes. (See Appendix C: Special Minority Jobs, Jobs, and Jobs.)

Many of our civil rights leaders have suggested for years that we minorities start writing books and documenting items of interest. This will help bring change to issues that need to be addressed and help provide jobs for minority contractors and minority youth. I have listened to our leaders, and I hope that my books will help overcome the struggles in minority contracting and the hiring of minority youth.

Most of my success during my rehabilitation was due to the young minorities in America and in the motherland of Africa in Kenya—they are very sharp and very smart and need to be given a chance. Please build up our country by giving the young and smart youth of today an opportunity to work, perform, and serve.

Remember that God gives, and God takes away. That is why I decided to be thankful, not vengeful, and let God deal with the unscrupulous and discriminating

"Wall Street Money Contractors." God helped me at just the right time, giving me the ability, because of a wink of my eye, to again walk, talk, and think. That, my good friends, is very powerful good news.

www.ingramcontent.com/pod-product-compliance
Lightning Source LLC
Chambersburg PA
CBHW031919240526
45464CB00021B/502